AF362027

Advances in the Ecohydrology of Arid Lands

Advances in the Ecohydrology of Arid Lands

Editors

Philip Micklin
Pingping Luo

MDPI • Basel • Beijing • Wuhan • Barcelona • Belgrade • Manchester • Tokyo • Cluj • Tianjin

Editors
Philip Micklin
Western Michigan University
USA

Pingping Luo
Chang'an University
China

Editorial Office
MDPI
St. Alban-Anlage 66
4052 Basel, Switzerland

This is a reprint of articles from the Special Issue published online in the open access journal *Hydrology* (ISSN 2306-5338) (available at: https://www.mdpi.com/journal/hydrology/special_issues/EAL).

For citation purposes, cite each article independently as indicated on the article page online and as indicated below:

LastName, A.A.; LastName, B.B.; LastName, C.C. Article Title. *Journal Name* **Year**, *Volume Number*, Page Range.

ISBN 978-3-0365-3323-0 (Hbk)
ISBN 978-3-0365-3324-7 (PDF)

Cover image courtesy of Philip Micklin.

Contents

About the Editors

Philip Micklin is Professor Emeritus of geography at Western Michigan University where he taught for 30 years before retirement in 1992. He has authored or coauthored more than 100 periodical articles and book chapters, two long monographs, as well as editing three books. Most of his research work has focused on water management issues in the former Soviet Union and the new nations of Central Asia. He has also written on global and U.S. water issues. Micklin has continued his publication activities in recent years, including as chief editor for a major book on the Aral Sea published in 2014 by Springer. He has also had field experience with applied hydrology. Dr. Micklin has spent considerable time in the Soviet Union, Russia and the new nations of Central Asia. He was given the WMU Distinguished Faculty Scholar award in 1992.

Pingping Luo is a full professor in School of Water and Environment, Chang'an University. He was graduated from Disaster Prevention Research Institute (DPRI), Kyoto University in 2012. He did postdoc researcher in DPRI, Kyoto University and IAS, Unitied Nations University from 2012 to 2016. Prof. Luo was appointed as full professor in Chang'an University from 2016. He has published more than 60 papers and 5 book chapters. His major is focused on Hydrological modelling, Flood disaster, Water resource management and Environmental Engineering.

Editorial

Editorial for Special Issue "Advances in the Ecohydrology of Arid Lands"

Philip P. Micklin [1],* and Pingping Luo [2]

1   Geography Department, Western Michigan University, 3219 Wood Hall, Kalamazoo, MI 49001, USA
2   School of Water and Environment, Chang'an University, Xi'an 710054, China; lpp@chd.edu.cn
*   Correspondence: philip.micklin@wmich.edu

Citation: Micklin, P.P.; Luo, P. Editorial for Special Issue "Advances in the Ecohydrology of Arid Lands". *Hydrology* **2022**, *9*, 33. https://doi.org/10.3390/hydrology9020033

Received: 24 January 2022
Accepted: 6 February 2022
Published: 16 February 2022

Publisher's Note: MDPI stays neutral with regard to jurisdictional claims in published maps and institutional affiliations.

Ecohydrology is an emerging, cross disciplinary subfield of hydrology devoted to the mutual interactions between water and ecosystems [1]. Today, the important questions of what these interactions mean for human society and how human society impacts these interactions are also part of this subject. The specific climatic/geographic focus here is arid lands broadly defined as water deficit regions where potential evapotranspiration (PET) exceeds precipitation (P). Such lands exceed 41% of the world's terrestrial area and are found on all continents except Antarctica [2]. Their range is from the climatic/vegetation classifications deserts and semi-deserts that are excessively dry most of the time through True Steppes to Wooded Steppes, Mediterranean areas and Tropical Savannas which suffer from moderately dry conditions parts of the year. Some mountainous regions and parts of polar lands also at times experience arid conditions.

Arid lands are of great contemporary concern as human-influenced climate change, considered by many experts as Humanity's greatest existential threat, is so strongly and negatively affecting them with, among other consequences, more frequent and severe drought and devastating fires. The experience of the state of California in the United States is a prime example of this, where, in recent years, economic losses from drought and fire ran into the tens of billions of dollars accompanied by significant human fatalities. Moreover, drought in northwestern China in recent years has caused ecological degradation and sandstorms. The monetary losses and deaths are expected to become higher as the climate warms. Ecohydrology has contributed to a better understanding and mitigation of these phenomena and will do so even more in the future.

An interesting and instructive example of the value of ecohydrology in better understanding and managing water in arid areas of the planet has been the effort to partially rehabilitate the northern part of the Aral Sea. The author of this editorial has been involved with research on this subject since the early 1980s, including on-site field research, data gathering and the publication of results.

The Aral Sea lies among the deserts and steppes of Central Asia. It was the world's fourth largest lake in surface area in 1960, but owing to human actions (primarily expanded irrigation in its basin) has steadily shrunk over ensuing decades with enormous negative ecological, economic and human welfare consequences [3]. Efforts were begun by the former Soviet Union in the late 1980s to address the problem, and after the breakup of the USSR in 1991, international organizations of the UN, the World Bank and various developed nations joined the effort. It rapidly became apparent that restoring the sea to its former size was impossible in any realistic near-term scenario, owing to costs and lack of the necessary water. However, a plan to partially revitalize the northern part of the lake was viewed as attainable as its level could be raised and stabilized by placing a regulating dam in the former strait separating the Small (north) Aral from the Large (south) Aral [3–5]. A locally constructed, closed-earthen dike had demonstrated the feasibility of the concept in the 1990s but catastrophically failed in 1999 as the water level behind the dam rose so high that the facility was breached during a major wind storm.

The World Bank, in collaboration with the Government of Kazakhstan, on whose territory the Small Aral is located, developed a full plan and engineering design by the early 2000s. The work involved flow measurements of the influent river (Syr Darya) as well as accompanying studies of and data gathering on the limnology, biology and fishery potential of the Small Aral Sea. A Russian hydroengineering company with considerable foreign experience built the key engineering facilities needed (Kok-Aral Dike and Dam) between 2003 and late 2005. The main component of the plan is a well-designed, engineeringly sound 13 km earthen dike with a concrete flow control dam with adjustable flow regulating gates to control outflow and the level of the Small Aral.

So far, the plan seems a considerable success. It raised the level of the Small Aral by two meters and increased the area by 18%. This led to a drop in salinity to levels characteristic of the pre-desiccation situation and allowed the return of native fish species to the lake from the Syr Darya Delta and its associated lakes. Fish catches have grown rapidly and allowed a thriving fish canning and export industry to develop in Aralsk—the former major port on the Small Aral [6]. Fresh and canned fish are supplied locally and to surrounding regions. One highly prized species, the Pike-perch (Sudak in Russian) is made into filets, frozen in local plants and exported to countries in the European Union, where it commands a high price. The revitalization of the Small Aral and its positive impacts on the lake's ecology and fisheries has been a boon to the local and regional economy and the welfare of the population living near the Small Aral.

The intent is to implement a second stage of the Small Aral Project. Two plans are being evaluated to further improve the ecology and fishery significance of the Small Aral. One would raise the level of one part of the Small Aral, and the other would raise the entire water body. Clearly, sufficient inflow is available for the former plan. The latter effort is a closer call, although a recent study seems to suggest it is feasible [4]. Ecohydrological research should play a key role in determining which variant is best, taking in to account not only physical but economic and social priorities.

The subfield of ecohydrology is changing rapidly. The intent of this SI is to present scientifically accurate information on the current state of leading ecohydrology-oriented research on arid lands that represents contemporary best thinking in the field. The five research articles included (and cited below) by no means cover the diversity of the field but rather provide an introduction to leading current research. The intended audience is broad, including not only those involved in this field but also those engaged in the more traditional aspects of hydrology, biology, ecology, geography, engineering, water management, agriculture, urban planning and other relevant fields.

Articles in the Special Issue

Ndiaye, P.; Bodian, A.; Diop, L.; Deme, A.; Dezetter, A.; Djaman, K. Evaluation and Calibration of Alternative Methods for Estimating Reference Evapotranspiration in the Senegal River Basin. Hydrology 2020, 7(2), 24; https://doi.org/10.3390/hydrology7020024. https://www.mdpi.com/2306-5338/7/2/24. Reference evapotranspiration (ETo) is a key element of the water cycle in tropical planning and management of water resources, hydrologic modeling, and irrigation management. This research assesses 20 methods for computing ETo.

Nabih, S.; Tzoraki, O.; Zanis, P.; Tsikerdekis, T.; Akritidis, D.; Kontogeorgos, I.; Benaabidate, L. Alteration of the Ecohydrological Status of the Intermittent Flow Rivers and Ephemeral Streams due to the Climate Change Impact (Case Study: Tsiknias River). Hydrology 2021, 8(1), 43; https://doi.org/10.3390/hydrology8010043. https://www.mdpi.com/2306-5338/8/1/43. Climate change projections predict the increase in no rain periods and storm intensity, resulting in a high alteration of Mediterranean rivers. Intermittent flow rivers and ephemeral streams, such as the Tsiknias, are especially vulnerable to spatiotemporal variation in climatic parameters, land use changes, and other anthropogenic factors.

Guerreiro, M.; Maia de Andrade, E.; Palácio, H.; Brasil, J.; Ribeiro Filho, J. Enhancing Ecosystem Services to Minimize Impact of Climate Variability in a Dry Tropical Forest with Vertisols. Hydrology 2021, 8(1), 46; https://doi.org/10.3390/hydrology8010046.

https://www.mdpi.com/2306-5338/8/1/46. Increased drought and variable rainfall patterns may alter the capacity to provide ecosystem services such as biomass production and clean water provision. The impact of these factors in a semi-arid region with dry tropical forests whose soils are mainly vertisols is the focus of this study.

Tsutsui, H.; Sawada, Y.; Onuma, K.; Ito, H.; Koike, T. Drought Monitoring over West Africa Based on an Ecohydrological Simulation (2003–2018). Hydrology 2021, 8(4), 155; https://doi.org/10.3390/hydrology8040155. https://www.mdpi.com/2306-5338/8/4/155. Food production in West Africa has sharply declined in recent years owing to agricultural drought. We simulated ecohydrological variables contributing to the drought using the Coupled Land and Plant hydraulics Model.

Frau, D.; Moran, B.; Arengo, F.; Marconi, P.; Battauz, Y.; Mora, C.; Manzo, R.; Mayora, G.; Boutt, D. Hydroclimatological Patterns and Limnological Characteristics of Unique Wetland Systems on the Argentine High Andean Plateau. Hydrology 2021, 8(4), 164; https://doi.org/10.3390/hydrology8040164; https://www.mdpi.com/2306-5338/8/4/164. We describe the hydroclimatological and limnological characteristics of 21 wetlands on the High Andean Plateau of Argentina, synthesizing information gathered over 10 years (2010–2020).

Author Contributions: The two guest editors equally shared the work of soliciting, editing, and handling the contributions to this special issue. All authors have read and agreed to the published version of the manuscript.

Funding: The creation of this special issue did not receive external funding.

Data Availability Statement: Not applicable.

Acknowledgments: We would like to acknowledge the efforts of all authors that added to the special issue.

Conflicts of Interest: The authors declare no conflict of interest.

References

1. Acreman, M.; Blake, J.; Carvalho, L.; Dunbar, M.; Gunn, I.; Gustard, A.; Jones, I.; Laize, C.; Maberly, S.; Mackay, E. Ecohydrology. In *Progress in Modern Hydrology: Past, Present and Future*; John, C.R., Robinson, M., Eds.; John Wiley & Sons, Incorporated. (ProQuest Ebook) Central: Hoboken, NJ, USA, 2015; Chapter 9; pp. 267–301. Available online: http://ebookcentral.proquest.com/lib/wmichlib-ebooks/detail.action?docID=4040936 (accessed on 5 March 2020).
2. D'Odorico, P.; Porporato, A. (Eds.) Ecohydrology of Arid and Semiarid Ecosystems: An Introduction. In *Dryland Ecohydrology*; Springer: Dordrecht, The Netherlands, 2006; pp. 1–11.
3. Micklin, P.; Aladin, N.; White, K.; Chida, T.; Boroffka, N.; Plotnikov, I.; Krivonogov, S. The Aral Sea: A Story of Devastation and Partial Recovery of a Large Lake. In *Large Asian GUnn in a Changing World*; Mischke, S., Ed.; Springer Nature: Cham, Switzerland, 2020; pp. 109–143.
4. Micklin, P. The future Aral Sea: Hope and despair. *Environ. Earth Sci.* **2016**, *75*, 1–15. [CrossRef]
5. Micklin, P. Efforts to Revive the Aral Sea. In *The Aral Sea: The Devastation and Partial Rehabilitation of a Great Lake*; Micklin, P., Aladin, N., Plotnikov, I., Eds.; Springer Earth System Sciences: Heidelberg, Germany, 2020; pp. 361–380.
6. White, K.; Micklin, P. Ecological restoration and economic recovery in Kazakhstan's Northern Aral Sea region. *Focus Geogr.* **2021**, *64*, 162. [CrossRef]

Article

Hydroclimatological Patterns and Limnological Characteristics of Unique Wetland Systems on the Argentine High Andean Plateau

Diego Frau [1,*], Brendan J. Moran [2,*], Felicity Arengo [3], Patricia Marconi [4], Yamila Battauz [5], Celeste Mora [1], Ramiro Manzo [1], Gisela Mayora [1] and David F. Boutt [2]

[1] Instituto Nacional de Limnología (CONICET-UNL), Ciudad Universitaria S/N, Santa Fe 3000, Argentina; celemora@hotmail.com (C.M.); ramiromanzo@gmail.com (R.M.); giselamayora@hotmail.com (G.M.)
[2] Department of Geosciences, University of Massachusetts-Amherst, 233 Morrill Science Center, 627 North Pleasant Street, Amherst, MA 01003-9297, USA; dboutt@geo.umass.edu
[3] Center for Biodiversity and Conservation, American Museum of Natural History, 200 Central Park West, New York, NY 10024-5102, USA; farengo@amnh.org
[4] Fundación YUCHAN, Mariano Moreno 1950, Villa San Lorenzo, Salta 4401, Argentina; huaico1709@gmail.com
[5] Facultad de Ciencias y Tecnología, Universidad Autónoma de Entre Ríos, Ruta Provincial N° 11. Km 10.5, Oro Verde 3100, Argentina; yamilabattauz@gmail.com
* Correspondence: diegofrau@gmail.com (D.F.); bmoran@umass.edu (B.J.M.)

Citation: Frau, D.; Moran, B.J.; Arengo, F.; Marconi, P.; Battauz, Y.; Mora, C.; Manzo, R.; Mayora, G.; Boutt, D.F. Hydroclimatological Patterns and Limnological Characteristics of Unique Wetland Systems on the Argentine High Andean Plateau. *Hydrology* **2021**, *8*, 164. https://doi.org/10.3390/hydrology8040164

Academic Editor: Mohammad Valipour

Received: 21 September 2021
Accepted: 30 October 2021
Published: 3 November 2021

Publisher's Note: MDPI stays neutral with regard to jurisdictional claims in published maps and institutional affiliations.

Abstract: High-elevation wetlands in South America are not well described despite their high sensitivity to human impact and unique biodiversity. We describe the hydroclimatological and limnological characteristics of 21 wetlands on the High Andean Plateau of Argentina, synthesizing information gathered over ten years (2010–2020). We collected physical-chemical, phytoplankton, and zooplankton data and counted flamingos in each wetland. We also conducted an extensive analysis of climatic patterns and hydrological responses since 1985. These wetlands are shallow, with a wide range of salinity (from fresh to brine), mostly alkaline, and are dominated by carbonate and gypsum deposits and sodium-chloride waters. They tend to have high nutrient concentrations. Plankton shows a low species richness and moderate to high dominance of taxa. Flamingos are highly dependent on the presence of Bacillariophyta, which appears to be positively linked to silica and soluble reactive phosphorus availability. Climatic conditions show a strong region-wide increase in average air temperature since the mid-1980s and a decrease in precipitation between 1985–1999 and 2000–2020. These high-elevation wetlands are fundamentally sensitive systems; therefore, having baseline information becomes imperative to understanding the impact of climatic changes and other human perturbations. This work attempts to advance the body of scientific knowledge of these unique wetland systems.

Keywords: high elevation wetlands; plankton; flamingos; hydroclimatic patterns; limnology; Andean mountains

1. Introduction

The dry Andean plateau stretches 1800 km along the backbone of the mountain range from southern Peru to northern Argentina between the eastern and western mountains, varies from 350 to 400 km in width and averages 4000 m above sea level (m.a.s.l) [1,2]. Despite the aridity, the plateau is dotted with wetlands that provide essential resources for human activity and habitats for biodiversity highly adapted to extreme environmental conditions [3,4]. Nevertheless, these wetlands have received less attention than other types of water systems on the South American continent. This is likely linked to the difficulty of obtaining samples, the variety of morphological and hydrological characteristics, the complexity of the hydrogeology, and the abundance and variety of basins and water bodies in this environment [3,5].

The isolated wetlands of the Andean plateau are formed in arid and semiarid climates where annual evaporation exceeds precipitation, forming endorheic drainage basins with deep groundwater flow systems and little surface water [5,6]. This results in the development of unique wetlands at their terminal areas where salt concentrations may reach 250 g L^{-1}, and evaporite deposits called *salars* often form [7,8]. Due to the extreme environmental conditions (high elevation, low oxygen availability, high irradiance, high desiccation potential, salinity gradients, and large daily air temperature fluctuations causing wetlands to freeze during winter), high-elevation wetlands tend to have simple food webs and react more rapidly and more sensitively partially or entirely to environmental changes than wetlands at lower elevations [9,10]. Vegetation patches (peatlands or high-elevation bogs), known locally as *vegas* and *bofedales*, are commonly found at the perimeter of lagoons and play a critical role in regulating the local water balance and sustaining a unique diversity of rare and endemic biota in the high Andes [11,12].

Regarding the geology and hydrology of these arid, endorheic landscapes, the difficulty of constraining fundamental hydrological processes such as response times, flow paths, and distribution and timing of groundwater recharge is magnified by long residence times (>100 years), deep water tables (>100 m), and often insufficient data [13,14]. Indeed, Liu et al. [15] showed that one in three of these catchments had an effective catchment area larger or smaller than the topographic catchment, leading to large discrepancies in water budgets. These hydrological imbalances within a catchment may be reconciled by the subsurface interbasin flow between topographic drainages and/or the draining of stored groundwater recharged during wetter periods hundreds to thousands of years before the present [16]. These processes are well documented globally [13,16–18]; and particularly in the Chilean pre-Andean depression, where relevant contributions have been made recently [19–22]. As in other arid regions of South America, these kinds of studies are relatively scarce in Argentina. Many questions remain to be answered, such as the overall response to hydroclimate changes at the higher elevation wetlands and slightly wetter climates on the plateau. Specifically, we still have insufficient knowledge about the reaction of vegetation extent in response to observed climate changes and how these responses vary by wetland elevation, basin area, and relative water age feeding surface water features.

In addition to the natural climatological variability, the wetland systems in the study area have been influenced by human activity to some extent since humans first settled here. However, these systems are the main providers of ecosystem services in a region where water is a scarce and limiting resource [23]. Many of the salt lake brines and flats of the Andean plateau are rich in sodium, lithium, boron, magnesium, and other elements that attract mining. Indeed, mining has been present here throughout history, from small-scale mines to major national and multinational investments [24]. Some of the largest lithium mineral deposits in the world lie within or near these wetland systems [25,26], and lithium mining activity has increased in size and scope since 2016 to supply the growing markets for rechargeable batteries. Changes in land use, such as agriculture expansion, nutrient enrichment, organic pollution, and an increased heavy metal load, also have a growing impact on these kinds of wetland ecosystems [27,28]. In addition to the local and regional pressures, global climate change is causing significant increases in temperature and changes in precipitation patterns which will likely affect the size and distribution of glaciers and wetlands, ecosystem integrity, surface water extent, and water availability for human activities [29–32].

The wetlands of the high Andes have also been identified as important ecoregions of diversity with high levels of endemism, unique traits, and evolutionary novelty [33]. Several rich, polyextremophilic microbial ecosystems flourish in these extreme environments. Biofilms, endoevaporitic mats, domes, and microbialites have been found to exist in association with salars, lagoons, and even volcanic fumaroles in central Andean environments, e.g., [34,35], for a complete review. Regarding planktonic microorganisms, several valuable contributions to the knowledge of the plankton diversity of high elevation wetlands have been published [3,36–43] with some other relevant contributions such as Servant-Vildary and Roux [44], Rejas et al. [45], and Frau et al. [46]. These authors also demonstrated the importance of nutrient limitation, especially silica and ionic composition for high-elevation diatom assemblages.

The high Andean wetlands support a diversity of endemic, resident, and migratory birds, and some of them completely dependent on water [47,48]. These wetlands are one of the main drivers of bird spatial distribution at high elevation and have a crucial role in preserving birdlife [49,50]. They also serve as steppingstones during the migration of long-distance migrants, particularly shorebirds [48,51,52]. Among birds, flamingos (*Phoenicopteridae*) are iconic species in the landscape, moving from one wetland to another, making alternative and complimentary use of these wetlands for feeding and breeding, effectively connecting these wetlands as a network of habitats [53–55]. Flamingos are an important component of the increasing nature tourism industry that has developed in the past decade and serve as flagship species for the conservation of high-elevation watersheds. Native mammals that depend on these wetlands to survive include puma (*Puma concolor*), Andean cat (*Leopardus jacobitus*), Andean fox (*Lycalopex culpaeus*), hairy armadillo (*Chaetophractus nationi*), and the commercially valuable vicuña (*Vicugna vicugna*) [56].

Despite all the singularities reported above regarding high-elevation wetlands from the Andean Plateau, human activities in combination with global climate change are driving variations in biological and hydrological attributes to a rate faster than we can measure. Baseline information regarding biological conditions at these wetlands, highly sensitive to anthropogenic disturbances [9] and, thus, ideal sentinels of global change [57,58] are becoming necessary to improve our capacity to properly manage the use of these environments [30,59,60]. Catamarca province, located in northwest Argentina, is an area where human activity, mainly mining has increased recently [61]. In Catamarca, we have gathered information over the last ten years to characterize hydroclimatological and limnological aspects, but we lack a holistic perspective of the characteristics and functioning of these systems. We argue that assessing geochemical and limnological characteristics at specific wetland systems in Catamarca can significantly improve the understanding of potential impacts from current and future human activities in these and similar systems across the High Andean Plateau. High-elevation wetlands are considered one of the most comparable ecosystems across continents [57] due to the many ecological features that they share. This makes the information synthesized in this manuscript potentially applicable to many other environments in South America and other arid regions that have not been adequately studied but share similar issues of human modifications and climatic change. In this study, we summarize the findings of a multidisciplinary approach synthesizing information gathered from publicly available data sources and field expeditions, describing the hydrogeological components and relevant biological patterns of wetlands on the High Andean Plateau in northwest Argentina.

2. Methods

2.1. Study Area

The study area encompassed 21 wetlands sampled in northwestern Argentina between 25°–28° S and 67°–69° W and situated between 3000 and 4500 m above sea level (m.a.s.l) (Figure 1). The area is characterized by predominantly volcanic geomorphology with numerous stratovolcanoes and vast basaltic and pyroclastic deposits. Eight of the wetlands sampled are within the *Lagunas Altoandinas y Puneñas de Catamarca* Ramsar site, which was designated in 2009. The climate is cold and dry, with temperatures below 0 °C most of the year and average annual rainfall between 50 and 200 mm year^{-1} [62]. Most of these wetlands are shallow, commonly less than 1 m in depth, and become partially or entirely frozen during winter [10]. They all sit within endorheic basins where water inflow comes mainly from groundwater recharge in the surrounding mountains and local alluvial infiltration and runoff, all of which leave the system through evapotranspiration.

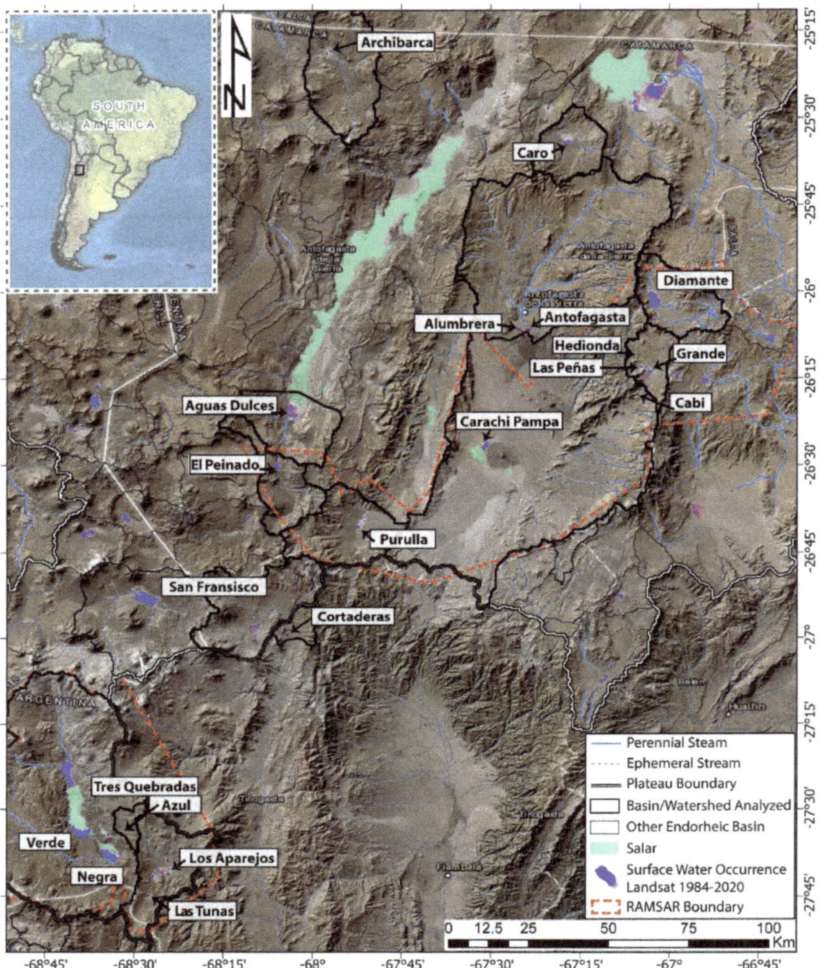

Figure 1. Map of the High Andean Plateau of Argentina and major hydrological features. The wetlands in this study are labeled, and the watersheds/basins for each wetland complex are outlined. The surface water occurrence layer depicts the average occurrence of water over the Landsat record on a scale from pale purple meaning very little occurrence (<10%) to dark blue meaning permanent water (100%). This dataset is publicly available through the European Commission's Joint Research Centre [63].

2.2. Physical-Chemical Sampling and Laboratory Analyses

Sampling was conducted between 2010 and 2020 during the summer season (January–February) because some of these wetlands are dry or frozen during the rest of the year, over seven consecutive days in a total of 21 wetlands, including peatlands, lagoons, and salars (see Table S1 in the Supplementary Materials for specific sampling dates at each site). At each sampling site, between one and four replicates were taken for the several hydrogeochemical-limnological parameters explained below. It is important to note that our assessment included 21 wetlands in total; however, no biological or limnological data was collected at Laguna Diamante, meaning there are only 20 wetlands included in those results.

For limnological analyses, measurements made in situ were temperature ($^{\circ}$C), pH, and conductivity (mS cm^{-1}) using HANNA multiparametric meters; transparency with a Secchi disc (m); and elevation (m.a.s.l) using a GPS altimeter and verified with a digital elevation model in ArcGIS. Water-type categories were derived from the fresh-brackish-saline-brine categorization scheme described by Warren [64].

Water samples were taken for determination of the nutrients nitrate+nitrite (N-NO$_3^-$+N-NO$_2^-$), ammonium (N-NH$_4^+$), soluble reactive phosphorus (SRP), and dissolved silica (Si). Samples for determination of N forms and SRP were immediately filtered through membrane filters (0.45 μm pore size). Subsequently, samples for nitrate + nitrite and ammonium determination were acidified to a pH < 2 with concentrated sulfuric acid, whereas samples for SRP determination were preserved by adding 5 mg L^{-1} of HgCl$_2$. All samples were stored in cold and darkness before transporting to the laboratory for analysis. Acidified water samples were neutralized in the laboratory. Samples used for determination of N-NO$_3^-$+N-NO$_2^-$ were estimated by reducing N-NO$_3^-$ with hydrazine sulfate and subsequent colorimetric determination of N-NO$_2^-$ [65]. N-NH$_4^+$ was determined using the indophenol blue method, SRP using the ascorbic acid method, and Si using the molibdosilicate method. Dissolved inorganic nitrogen (DIN) was calculated as the sum of N-NO$_3^-$+N-NO$_2^-$ and N-NH$_4^+$. The methodology proposed in APHA [66] was followed throughout, and nutrient concentrations were expressed as mg L^{-1} for silica and μg L^{-1} for DIN and SRP.

Physical and chemical data were explored by using principal component analysis (PCA). All data were log-transformed, except pH, and values obtained were standardized. Neighbor-joining clustering analysis was performed on the physical-chemical data obtained to find common patterns among wetlands sampled.

Water samples analyzed for major ion composition were collected at 10 of the 21 wetlands in February of 2019 and 2020 using a consistent, standardized procedure and repeated sampling from the same location when possible (Supplementary Materials, Table S2). Water samples were filtered through 0.45 μm filters using a plastic 60 mL syringe and were stored in clean HDPE bottles. The concentration of major elements was analyzed using inductively coupled plasma optical emission spectroscopy for major elements (Agilent 5110 ICP-OES) and high-pressure ion chromatography (Dionex Integrion HPIC) for Cl, SO$_4$, PO$_4$, and NO$_4$ anions at the University of Massachusetts Amherst. Waters with relatively high TDS were diluted volumetrically before analysis. For ICP OES analysis, samples were acidified to 1% HNO$_3$ *v/v* before analysis. Quantification was performed using seven external calibration standards ranging from 0.1 to 100 ppb. Calibration verification standards and blanks were run at every 10th analysis for anions and elements. Elemental analysis was verified with external NIST standard SRM 1643d, and anion analysis was verified with a secondary anion standard (Anion II Std Dionex). Samples that exceeded the calibration by 120% were diluted and reanalyzed. The concentration of HCO$_3$ in water was determined by titration using an autotitrator. In all cases, the laboratory measurement error was 0.01 mg L^{-1}. All other major ion data used in the study were obtained from Sureda [67]. A charge balance assessment of all these data was performed, and only samples with less than 10% error were included.

2.3. Hydroclimate Characterization

We utilized multiple remotely sensed datasets to assess hydrological and climatological conditions and changes over time across the basins of interest. These include Landsat satellite imagery (spatial resolution of 30 m $\times$ 30 m), the TerraClimate climatically aided precipitation interpolation, and the famine early warning systems network land data assimilation system (FLDAS). These data products are publicly available on Google Earth Engine (GEE) platform and were extracted and compiled using tools within the platform. For each wetland system, basin domains were extracted from the HydroSHEDS dataset (Hydrological data and maps based on shuttle elevation derivatives at multiple scales) [68] or a contributing watershed area determined using the Watershed tool in ArcGIS Pro.

Values of total precipitation and mean air temperature were extracted for each satellite pass (~16-day intervals) as spatially aggregated values for each basin or watershed containing the wetland of interest; outlines of these areas are shown in Figure 1. Total living or "green" vegetation extent was also assessed at each wetland complex using the normalized difference vegetation index (NDVI). These data provide a robust assessment of the change in climatic and hydrological conditions from the mid-1980s to the present.

To assess changes in precipitation, we utilized the TerraClimate precipitation product which uses climatically aided interpolation, combining high-spatial-resolution climatological normals from the WorldClim dataset, with coarser spatial resolution but time-varying data from CRU Ts4.0 and the Japanese 55-year reanalysis (JRA55) [69]. The data are provided as monthly precipitation accumulation totals at a resolution of $1/24°$ (~4-km) from 1958 to 2021. A time series of annual total precipitation was then created for each basin/watershed. From this time series, we derived a general value of precipitation changes between 1985 and 2020 represented by the percent change from the 1985–1999 mean value to the mean value of the 2000–2020 period.

Mean annual air temperature was determined using the famine early warning systems network (FEWS NET) land data assimilation system (FLDAS) global land surface simulation that provides monthly values of near-surface air temperature at a $0.1° \times 0.1°$ resolution [70]. The model is forced by combining the modern-era retrospective analysis for research and applications version 2 (MERRA-2) data and climate hazards group infrared precipitation with station (CHIRPS) 6 hourly rainfall data that have been downscaled using the NASA Land Data Toolkit. Although the precipitation product used in FLDAS is not as reliable for this region, the air temperature data are quite consistent with station data and other datasets and are, therefore, determined to be robust estimates. The value of temperate change from 1985 to 2020 was derived from the linear trend of mean annual air temperature over that period.

To assess changes in vegetation cover, we utilized the NDVI, which is calculated from spectral imagery using the formula: NDVI = (NIR − RED)/(NIR + RED) where NIR is the reflection in the near-infrared spectrum and RED is the reflection in the red range of the spectrum [71]. NDVI is designed to assess the density of vegetation at a given location at a given time. It provides a single band with a range from −1.0 to 1.0 where negative values are clouds, water, and snow; values close to 0 up to 0.1 are from rocks and bare soil. Different types of vegetation are classified in values greater than ~0.1. To be conservative and be sure we are capturing only vegetation, we extracted the pixels whose values fell between 0.2 and 0.9. The domains over which each wetland was analyzed were chosen to encompass all vegetation directly associated with the wetland system, at or near each basin floor. Using GEE, we extracted a full series of Landsat 5 and Landsat 7 Satellite images from 1985 to 2020 and determined the number of pixels covered by green vegetation (0.2–0.9 range) from which a total geographic area was then calculated. This provides a time series of the total area covered by living vegetation within each wetland region. From these time series, we derived a general value of change in the vegetated area between 1985 and 2020 represented by the percent change from the 1985–1999 mean value to the mean value of the 2000–2020 period.

2.4. Biological Sampling and Analyses

Samples of phytoplankton were collected at the same time as the physical-chemical samples from the subsuperficial area using 120 mL bottles and then were immediately fixed with 1% acidified Lugol solution for preservation. Quantitative sample analyses were carried out following the Utermöhl method [72], and the density obtained was expressed ind mL^{-1} by accepting a counting error <20% [73]. For estimating zooplankton abundance, between 10 and 30 L of water were filtered through a conventional conical plankton net (50 μm) with a collector and a 2 L bucket. The collected material was fixed in situ with 10% formalin and dyed with erythrosine. The microzooplankton counts (Rotifera and nauplii) were carried out with a conventional optical microscope using Sedgwick-Rafter type chambers with a capacity of 1 mL. Macrozooplankton (Cladocera and Copepoda) counts were performed in a 5 mL Borogov-type counting chamber. Density was expressed as ind L^{-1}. Taxonomic identification was carried out to the lowest taxonomic possible level. Phytoplankton taxonomic classifications were based on Lee [74]. Taxonomic determinations were made following keys and specific bibliography of each algal group, such as Krammer and Lange-Bertalot [75] for Bacillariophyta, Tell and Conforti [76] for Euglenophyceae, Komárek and Fott [77] for Chlorophyceae, and Komárek and Anagnostidis [78,79] for Cyanobacteria, among other authors and recent revisions. Zooplankton taxonomic identifications were based on Ahlstrom [80], Koste [81], Kořínek [82,83], Korovchinski [84], and Alekseev [85], among others. Flamingos were identified to species level (*Phoenicoparrus andinus*, *P. jamesi*, and *Phoenicopterus chilensis*) using spotting scopes and binoculars and counted from established census points, consistent across years. For large flocks (>500) or groups dispersed throughout the wetland, numbers were estimated using methods described in Marconi [86]. For this study, we used total numbers per flamingo species for each wetland.

Flamingo and plankton abundances were compared among wetlands by using Kruskal–Wallis analyses and the Mann–Whitney *post hoc* test. Temporal tendencies for flamingo abundances were explored by using Mann–Kendall tests using flamingo counts carried out from 2010 to 2020. Diversity indices, such as Shannon–Wieber, dominance, and richness, were estimated for plankton. Finally, a redundancy analysis (RDA) was performed by using all the environmental variables as predictors and the biological communities (flamingos, phytoplankton, and zooplankton) as response variables to explore the main environmental variables which may influence natural communities' abundance. The RDA analysis was performed after considering the longitude gradient in a detrended correspondence analysis (DCA), and highly correlated variables (VIF > 10) were omitted.

3. Results

3.1. Physical-Chemical and Hydroclimate Characterization

All the wetlands sampled (n = 21) were located above the 3000 m.a.s.l. and range in area from 24 to 1117 ha (determined by surface water extent from the HydroLAKES dataset described by Messager et al. [87] (Table 1). Diamante is not included in Table 1 since we did not collect biological or limnological data there; however, it is one of the largest wetlands in the region and therefore was included in the rest of our analysis. All of them were shallow (<1 m) with high transparency, allowing light to reach the bottom. Alkaline conditions were typical, except for Aguas Dulces showing to be circumneutral (pH = 7). Nutrient concentrations were variable across wetlands. DIN concentrations were above 100 µg L^{-1}, with the highest concentrations registered in Las Peñas, Carachi Pampa, Grande, and Hedionda and the lowest in Cortaderas, San Francisco, and Archibarca. SRP concentrations followed a similar pattern of high variation among lakes, with the highest concentrations registered in Archibarca and Hedionda. Values registered were above 50 µg L^{-1}, except in Antofagasta, Azul, and Negra (32.7, 44.1, and 46.1 µg L^{-1}, respectively). The DIN:SRP ratio showed, in general, low values (<10), indicating DIN limitation in almost all wetlands where information was available. Si concentrations were about one thousand times higher than concentrations of the other inorganic nutrients sampled (>30 mg L^{-1} for all lakes sampled). Conductivity showed a wide range, with some wetlands having very low values (e.g., Alumbrera, Antofagasta, and Cortaderas had <2.5 mS cm^{-1}). About 30% of the wetlands sampled were categorized as freshwater or brackish, all measuring below 10 mS cm^{-1}. The remaining wetlands (70%) were considered saline, with mean conductivity values between 25 and 110 (mS cm^{-1}) being the most saline; Diamante was a large lake composed of brine. (Table 1).

Table 1. Median environmental values for each wetland sampled between 2010 and 2020. Elevation (Alt), area (Area), temperature (Temp), pH (pH), water level (WL), Secchi disc (SD), conductivity (Cond), dissolved inorganic nitrogen (DIN), silica (Silica), soluble reactive phosphorus (SRP). ND = no data.

Wetlands	Type	Elev (m.a.s.l)	Area (ha)	Temp (°C)	pH	WL (cm)	SD (cm)	Cond (mS cm^{-1})	DIN (µg L^{-1})	Silica (mg L^{-1})	SRP (µg L^{-1})
Aguas Dulces	Saline	3325	543	26.2	7.0	5.0	5.0	25.6	ND	ND	ND
Alumbrera	Fresh	3320	24	19.8	9.7	3.1	3.5	2.2	227.5	40.7	77.0
Antofagasta	Fresh	3320	37.5	22.9	9.5	4.0	3.5	1.6	159.9	37.7	32.7
Los Aparejos	Saline	4220	95	10.7	9.0	3.0	3.0	74.0	146.8	29.5	70.1
Archibarca	Saline	4030	49.8	16.1	8.6	16.0	5.3	50.0	83.5	62.5	1962.8
Azul	Brackish	4450	66	7.8	9.5	7.0	7.0	3.9	156.4	10.2	44.1
Cabi	Saline	4220	67	23.5	9.5	2.0	2.0	51.5	ND	ND	ND
Carachi Pampa	Saline	3000	334	24.0	8.6	6.5	6.5	100.0	675.6	31.7	216.5
Caro	Saline	3990	511	19.8	8.1	1.0	1.0	73.8	ND	ND	ND
Cortaderas	Fresh	3900	51.01	22.5	8.7	18.8	18.8	1.3	53.6	70.3	142.0
Grande	Saline	4240	357	15.3	9.8	5.2	3.0	17.5	578.5	96.5	187.1
Hedionda	Saline	4280	20	25.1	9.8	3.0	3.0	38.6	370.3	206.7	3580.7
Negra	Saline	4090	1106	21.4	8.3	2.0	2.0	30.5	216.6	53.3	46.1
El Peinado	Saline	3750	241	17.0	9.0	28.2	5.0	67.3	ND	ND	ND
Las Peñas	Brackish	4255	53	12.4	9.0	6.9	4.8	6.0	1135.9	61.5	697.1
Purulla	Saline	3805	487.4	22.8	9.0	7.8	7.8	32.5	144.7	39.3	97.7
San Francisco	Brackish	4000	1066	12.0	8.2	12.9	11.5	6.8	57.9	73.7	966.8
Tres Quebradas	Saline	4090	ND	15.4	ND	6.0	6.0	108.6	ND	ND	ND
Las Tunas	Saline	4250	42	19.0	9.5	ND	ND	52.9	327.7	ND	51.9
Verde	Saline	4090	1077	17.2	ND	15	15	101.2	ND	ND	ND

The PCA ordination analysis explained 49.66% of the total variance in the first two axes. The first axis indicated a gradient of DIN concentration, with maximum values recorded in Carachi Pampa, Grande, and Las Peñas and the lowest value in Cortaderas. The second axis suggested a gradient of SRP concentration and elevation from San Francisco and Archibarca to Alumbrera and Antofagasta. Carachi Pampa, Negra, and Grande stand out for their high conductivity values, Si concentration, and area (Figure 2). Complementarily, the neighbor-joining clustering analysis suggests four clusters of wetlands. Cluster one (Archibarca, Cabi, Caro, Hedionda, El Peinado, and Las Peñas) had the highest nutrient values, being mostly saline lagoons. Cluster two (Aguas Dulces, Antofagasta, Alumbrera, Carachi Pampa, and Purulla) were those located at the lowest elevation (above 3000 m.a.s.l), had mean temperature values higher than the other wetlands (above 22 °C), and were saline-freshwater lakes. Cluster three (Grande, Verde, Negra, Tres Quebradas, and San Francisco) had the highest elevation, area, and Si concentrations and were saline-brackish. Finally, cluster four (Los Aparejos, Azul, Cortaderas, and Las Tunas) was characterized by having the lowest mean DIN and Si concentration, as was mostly saline-brackish (Figure 3).

Based on the dissolved ion compositions of 14 wetland surface waters (10 analyzed for this work and 4 from data obtained from Sureda [65], we can describe three distinct groups of wetlands (Figure 4). Group one: Grande, Diamante, Carachi Pampa, Caro, and Cabi are dominated by Na^+, K^+, and Cl^- with very low concentrations of Ca^{++}, Mg^{++}, and SO_4^-. Group two: Purulla, El Peinado, Negra, Verde, and San Francisco are also marked by high relative concentrations of Cl^- but have higher Ca^{++} and Mg^{++} concentrations than the first group. The third group of wetlands is marked by substantially higher relative concentrations of HCO_3 and SO_4^-, and include Alumbrera, Antofagasta, Azul, and Los Aparejos. Three of these are the freshest wetlands sampled (<5 mS cm^{-1}), and the other is much saltier (~75 mS cm^{-1}). Alumbrera and Antofagasta, which comprise the extensive wetland complex at the terminus of the large Punilla River, show very similar characteristics, being dominated by relatively high HCO_3 and SO_4^- but lower relative Cl^-, Ca^{++}, and Mg^{++} concentrations. Azul is like these lagoons but contains higher SO_4^- and much higher Mg^{++}. Los Aparejos appears to be more like group two than the fresher lagoons mentioned above but substantially lower in SO_4^-.

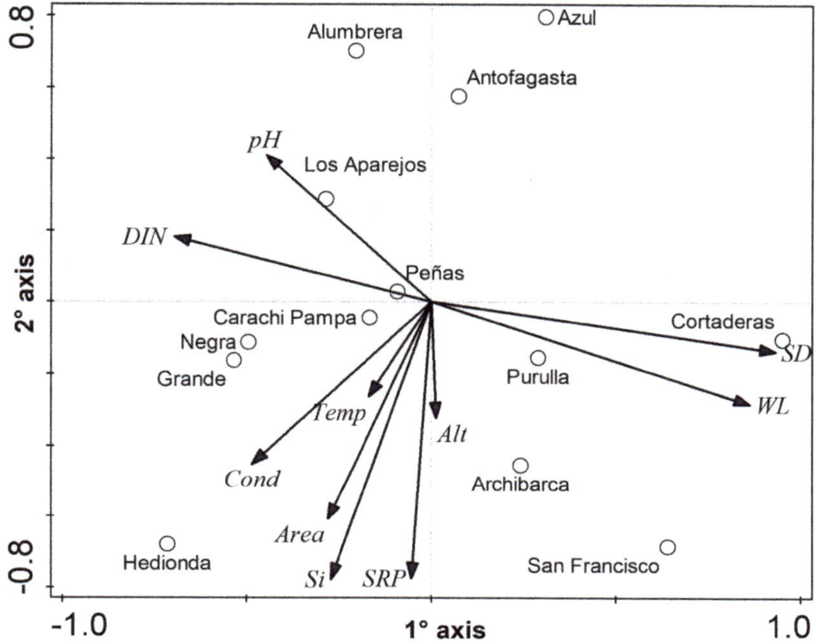

Figure 2. Principal components analysis (PCA) considering only those wetlands where full environmental data were available. Abbreviations of variable names as in Table 1.

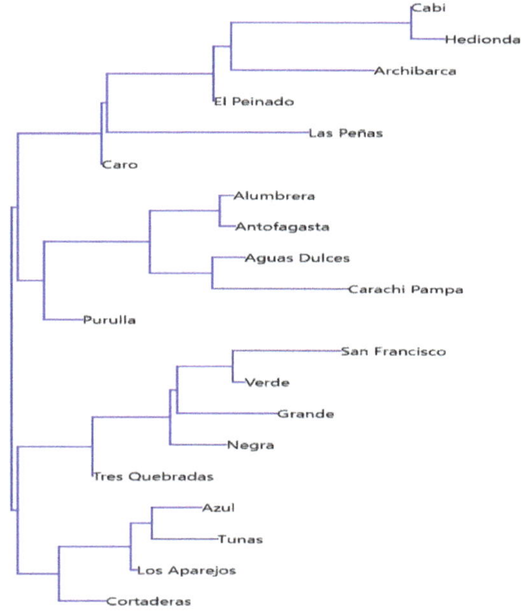

Figure 3. Neighbor-joining clustering analysis by using all the environmental variables (including nutrients) available.

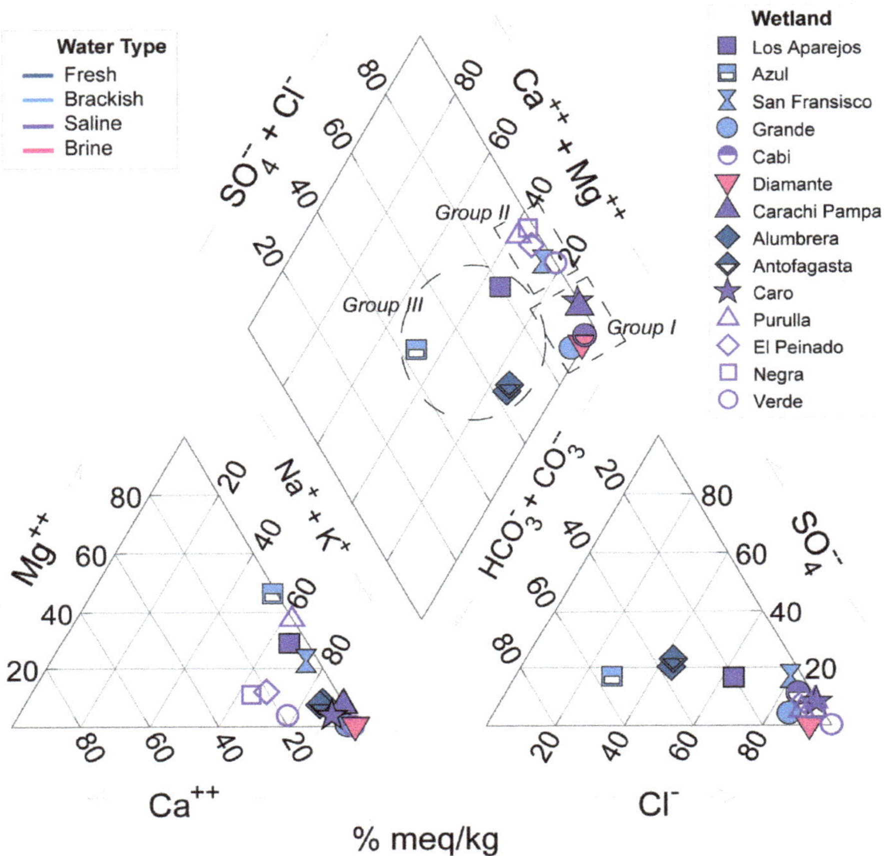

Figure 4. Piper diagram showing major ion compositions of wetland surface waters. Purulla, El Peinado, Negra, and Verde were data extracted from data presented by Sureda (2003).

Our analysis of climatic and hydrological conditions in this region revealed 14 distinct basins or watersheds encompassing the 21 wetlands in this study (Figure 5). Decadal-scale changes in climatic and hydrological conditions within these large internally drained basins provide an effective estimate of the natural drivers of recent hydrological changes in these wetland systems. In the case of Azul, Cortaderas, and Aguas Dulces, the domains used are subwatersheds of larger basins, and in the case of Cabi, a small wetland on the edge of the Grande Basin, its watershed area is too small to assess independently; therefore, we use the Grande basin as a representative of the conditions there. Temperatures since 1985 have increased substantially across the entire region. Every basin has recorded a temperature increase of at least 0.45 °C. Basins in the eastern and northern parts of the study area have seen an increase of between 0.55 and 0.65 °C with the largest increase observed in Archibarca and Caro (Table 2). Precipitation has broadly decreased across all wetland areas, with the smallest decreases observed in the northeastern region of between 9% and 12%. The most significant decreases are in the southern and westernmost regions of between 14% and 22%. Since a major region-wide drought has been identified since the mid-2000s, these changes are likely in large part a reflection of the relative depth of this drought across the study area. In terms of hydrological changes, in this case, represented by the change in vegetation cover, there is a wide distribution of relative changes. The western and southern wetlands (El Peinado, Tres Quebradas, Negra, Azul, and Las Tunas) as well as the highest elevation wetland, Diamante show large increases in total vegetation cover since the 1980s and 1990s with a range of between 12% and 53%, with the highest values in the Tres Quebrada basin. In contrast to this, many of the wetlands show a substantial decrease in vegetation cover from −10% to −31%. Particularly notable decreases were observed in the northern and eastern wetlands, where Archibarca, Antofagasta, Hedionda, Grande, and Cabi as well as the central basins San Francisco and Aguas Dulces all registered decreases in the vegetated area of >15%.

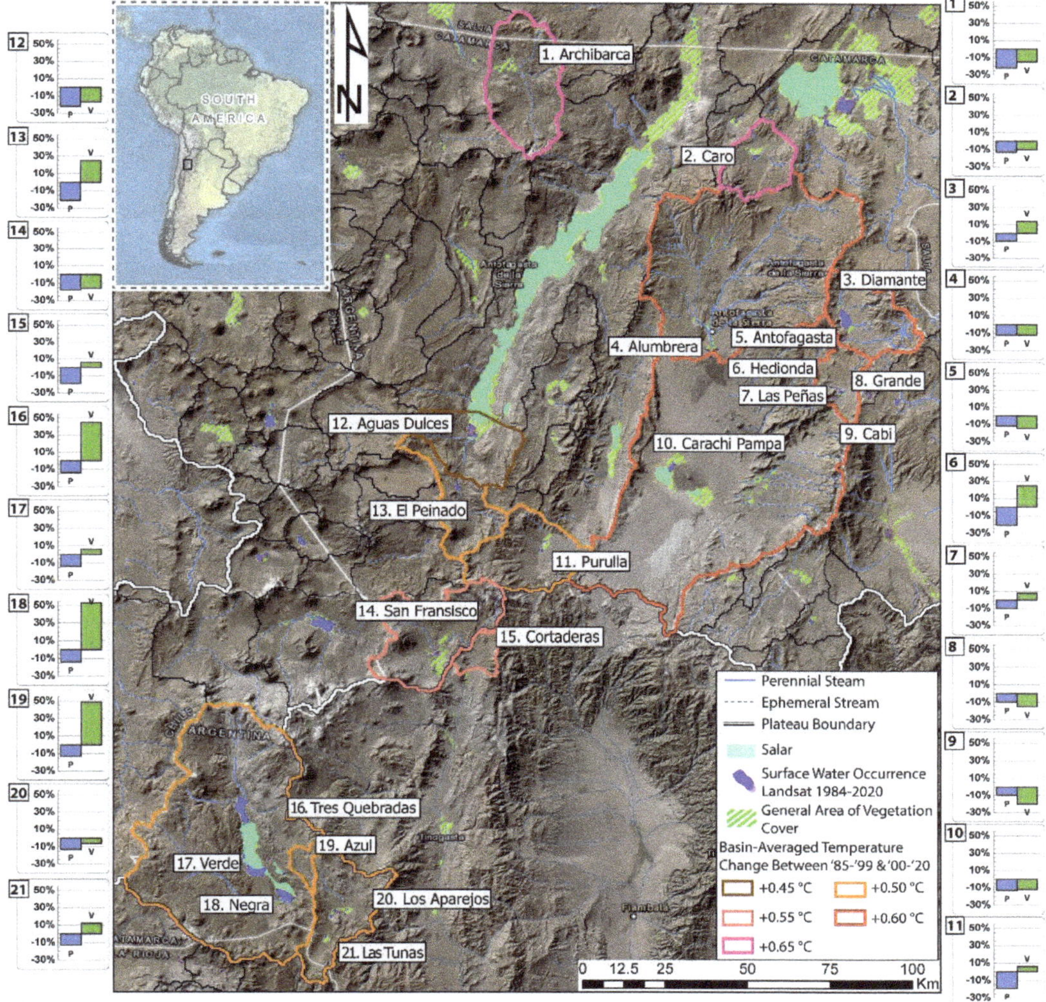

Figure 5. Map of recent hydroclimatological changes across the region relative to the 1985–99 period. Vegetation coverage (green hatch) is derived from the Landsat imagery record that outlines the regions of each wetland complex included in our analysis. Temperature changes observed in each basin of interest are outlined on a color scale by magnitude of change. Each wetland of interest (numbered) is labeled, and its corresponding subplot shows the observed changes in the relative vegetation extent and precipitation amount.

Table 2. Summary of hydroclimate results for each basin/watershed and associated wetland system. The largest observed decreases in precipitation, vegetation, and air temperature are highlighted in red, and the largest increases in vegetation extent are highlighted in green. Notes: [*1] Whole topographic basin minus the Antofasta/Alumbrera Watershed area. [*2] Based on linear trend.

Wetland	Basin/Watershed	Elevation (m.a.s.l)	Catchment-Wide Hydroclimate for 1985–1999 and 2000–2020 Periods		
			Precipitation Change	NDVI	Air Temp. Change (°C) [*2]
Archibarca	Archibarca Basin	4030	−22%	−15%	+0.65
Caro	Caro Basin	3990	−13%	−10%	+0.65
Diamante	Diamante Basin	4570	−9%	+14%	+0.60
Alumbrera	Antofasta/Alumbrera Watershed	3320	−12%	−12%	+0.60
Antofagasta	Antofasta/Alumbrera Watershed	3320	−12%	−15%	+0.60
Hedionda	Grande Basin	4280	−10%	−31%	+0.60
Las Peñas	Grande Basin	4255	−10%	+8%	+0.60
Grande	Grande Basin	4240	−10%	−15%	+0.60
Cabi	Grande Basin	4220	−10%	−20%	+0.60
Carachi Pampa	Carachi Pampa Basin [*1]	3000	−13%	−12%	+0.55
Purulla	Purulla Basin	3805	−19%	+6%	+0.50
Aguas Dulces	Aguas Dulces Watershed	3325	−22%	−17%	+0.45
El Peinado	El Peinado Basin	3750	−21%	+25%	+0.50
San Francisco	San Fransisco Basin	4000	−18%	−16%	+0.55
Cortaderas	Cortaderas Vega Watershed	3900	−19%	+6%	+0.55
Tres Quebradas	Tres Quebradas Basin	4090	−14%	+45%	+0.50
Verde	Tres Quebradas Basin	4090	−14%	+6%	+0.50
Negra	Tres Quebradas Basin	4090	−14%	+53%	+0.50
Azul	Azul Watershed	4450	−13%	+49%	+0.50
Los Aparejos	Los Aparejos Basin	4220	−13%	−6%	+0.50
Las Tunas	Los Aparejos Basin	4250	−13%	+12%	+0.50

3.2. Biological Communities' Characterization

Three flamingo species, *Phoenicoparrus andinus*, *P. jamesi*, and *Phoenicopterus chilensis*, were recorded in the study area during the sampling periods; however, their abundances were variable among wetlands and across years (Table 3). Flamingo abundance for all species among years was variable in some wetlands (e.g., San Francisco, Las Peñas, Aguas Dulces, and Archibarca) while appearing more stable in others (Carachi Pampa, Grande). *Phoenicoparrus jamesi* was the most abundant in Grande, where numbers were >12,000 in all years. This species was the most abundant in Los Aparejos, Hedionda, and Las Peñas. For *P. andinus*, numbers fluctuate less across years in San Francisco, Purulla, Negra, and Archibarca. *P. chilensis* numbers were very low in the study area considering a global population of >500,000 individuals. Las Peñas wetland consistently had *P. chilensis* present.

The Kruskal–Wallis test showed statistically significant differences among wetlands for the three species (H = 34.66 $p < 0.001$). These differences were for *P. jamesi* vs. *P. andinus* and *P. chilensis* and between *P. andinus* vs. *P. chilensis* (Mann–Whitney test, $p < 0.001$). When the temporal tendencies of the abundances between 2010 and 2020 were explored with the Mann–Kendall test, a statistically decreasing trend was detected for *P. chilensis* (Z = 3.42 $p < 0.0001$), notable since the year 2015. No tendency was seen for the other two flamingo species ($p > 0.05$) (Supplementary Material, Figure S1. Spearman correlation among all measured environmental variables and *P. jamesi* ($n = 13$) in wetlands with high variability among years (Alumbrera, Azul, Los Aparejos, and San Francisco) revealed a positive statistically significant correlation with Bacillariophyta (Rho = 0.71 $p = 0.006$), which showed a negative correlation with water level (Rho = −0.72 $p = 0.02$). A statistically significant positive correlation was found between *P. andinus* abundance and conductivity

(n = 11, Rho = 0.87 p = 0.01) which showed a marginally significant correlation with the area of the wetlands (Rho = 0.53 p = 0.06).

Table 3. Mean and percentage variation coefficient (%VC) of flamingos counted (individuals' numbers) per species in each wetland from 2010–2020.

Wetland	*P. andinus*		*P. jamesi*		*P. chilensis*	
	Mean	%VC	Mean	%VC	Mean	%VC
Aguas Dulces	52	63	1	100	14	50
Alumbrera	16	100	1	200	50	126
Antofagasta	9	144	1	200	77	222
Archibarca	248	37	83	128	52	92
Azul	6	167	2	150	5	100
Cabi	13	146	58	124	2	150
Carachi Pampa	91	25	31	71	10	120
Caro	62	84	42	138	17	47
Cortaderas	13	85	0	0	6	150
Diamante	158	47	163	142	65	63
Grande	161	88	14,988	18	19	126
Hedionda	37	65	277	48	12	167
Las Peñas	104	82	260	115	215	85
Las Tunas	69	113	131	109	12	108
Los Aparejos	96	77	376	69	23	148
Negra	182	49	12	83	4	100
El Peinado	179	26	4	200	84	63
Purulla	555	86	97	119	9	89
San Francisco	365	76	49	147	27	104
Tres Quebradas	8	88	0	0	0	0
Verde	0	0	1	100	0	0

For phytoplankton species, richness ranged between 3 and 20 species recorded for each wetland. Bacillariophyta species, especially from the genera *Navicula*, and *Nitzschia*, were the most frequently recorded. For Chlorophyceae, *Monoraphidium* and *Chlamydomonas* were the best-represented genera. For the other groups of algae, *Cryptomonas* (Cryptophyceae), *Lepocinclis*, *Euglena*, *Trachelomonas* (Euglenophyceae), and *Synechocystis*, *Oscillatoria*, *Phormidium* (Cyanobacteria) were the most representative. The assemblage tends to be dominated by Bacillariophyta, followed by Chlorophyceae and occasionally by Cyanobacteria. Density registered for phytoplankton was highly variable across wetlands with the highest values recorded in Grande, Cabi, and Las Peñas (maximum median values recorded = 18,635 ind mL^{-1}) and the lowest in Azul, El Peinado, and Tres Quebradas, with 107 ind mL^{-1} on average (Figure 6a). The high abundances found were attributed to Bacillariophyta which was the dominant group solely in Grande. The Shannon diversity index showed low values for these wetlands, always <3, and variable dominance values among wetlands (above 0.6 and 1). Alumbrera, Antofagasta, and Aguas Dulces reached the highest values and Hedionda and Las Tunas the lowest (Figure 6b). The Kruskal–Wallis analysis showed differences in total phytoplankton density among lakes (H = 30.49 p = 0.04), showing statistically significant differences between Grande and Antofagasta, Aparejos, and Azul (p < 0.05 in all cases).

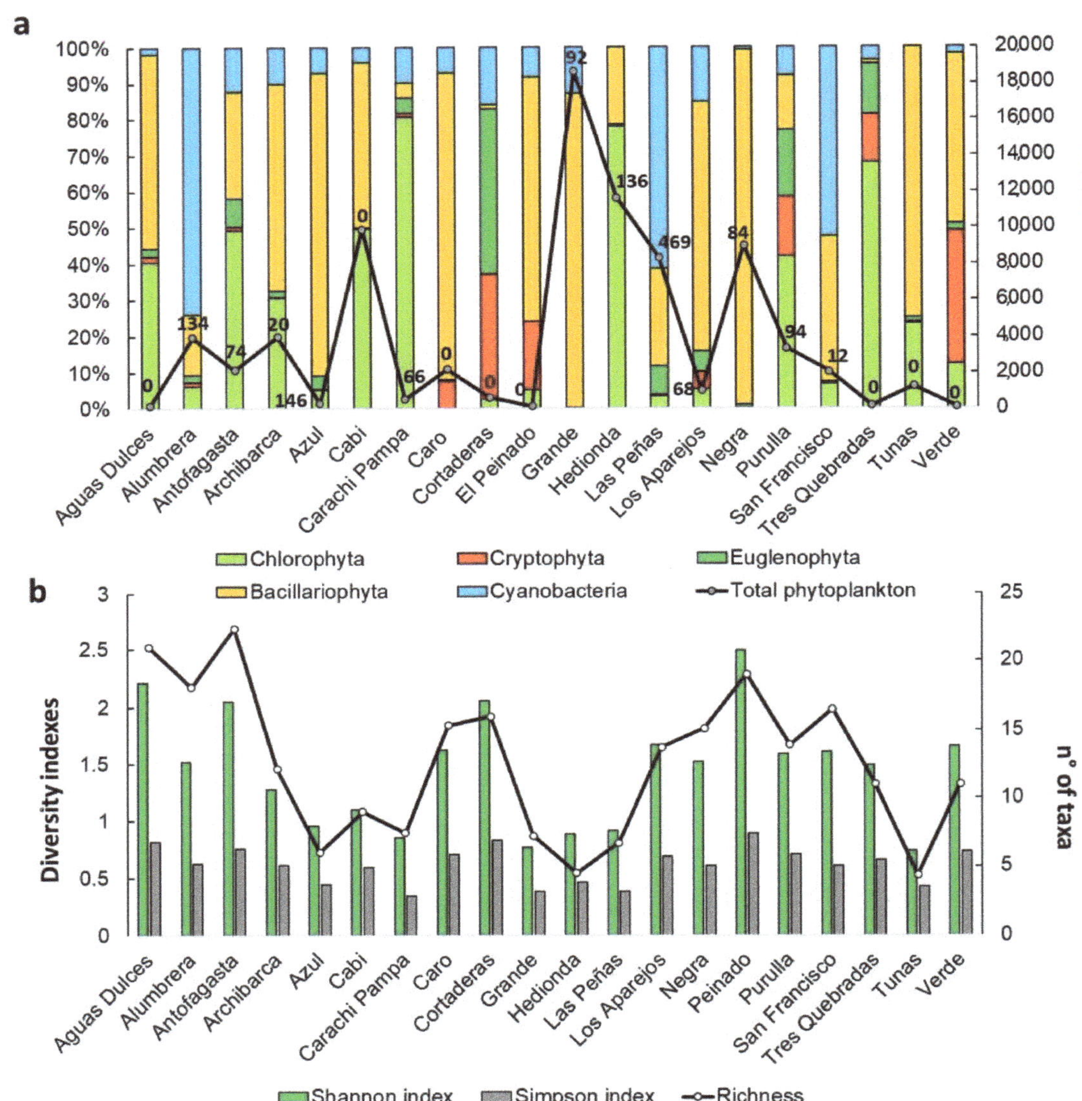

Figure 6. Phytoplankton relative percentage density registered for each group and total abundance (ind mL^{-1}) with percentage variation coefficient indicated (%VC) on each mean value (**a**) and diversity indexes (**b**) on each wetland where samples were obtained.

Zooplankton showed a similar pattern to phytoplankton, with between one and five species recorded for each lake and assemblages alternatively dominated by Copepoda, or Rotifera. *Alona* (Cladocera), *Lecane, Keratella, Brachionus,* and *Boeckella* (Copepoda) appeared as the most frequent genera. The density recorded for the different wetlands was, in general, low (median values of 10 ind L^{-1}), with Archibarca and Las Peñas being the exceptions (321 and 176 ind L^{-1}, respectively) (Figure 7a). The Shannon diversity index showed low values (<2) and highly variable values of dominance for the lakes sampled (between 0.2 and 1). Antofagasta, Alumbrera, and Azul had the highest richness values while Verde had the lowest (Figure 7b). The Kruskal–Wallis analysis showed no differences in density among wetlands (H = 20.37 p = 0.11).

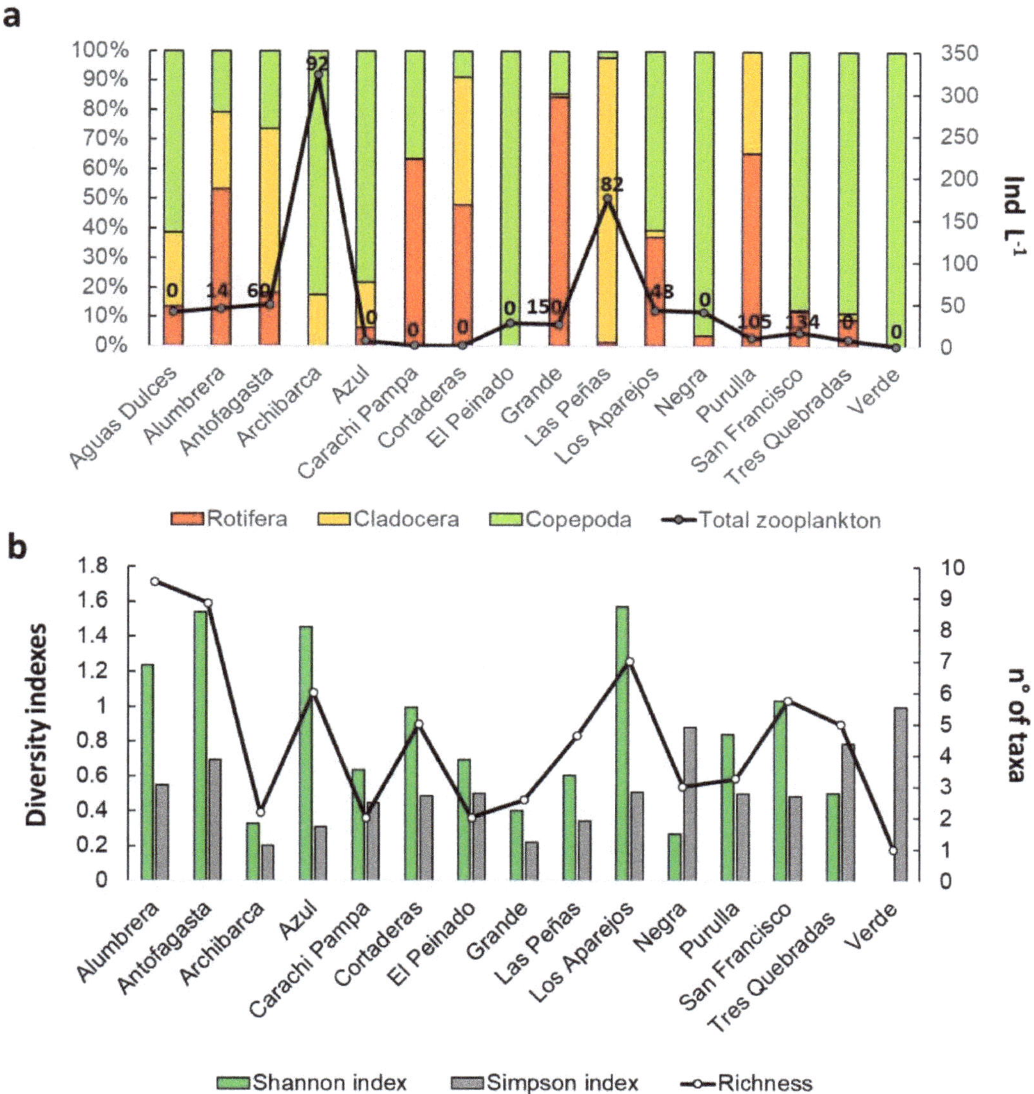

Figure 7. Zooplankton relative percentage density registered for each group and total abundance (ind L^{-1}) with percentage variation coefficient indicated (%VC) on each mean value (**a**) and diversity indexes (**b**) on each wetland where samples were obtained.

The RDA analysis explained 29.85% of the total variance (F = 1.8 p = 0.01), with the first two axes retaining 71.5% of the total variation. Both *P. andinus* and *P. jamesi* were associated with the highest elevation and area wetlands with higher abundances of Bacillariophyta, particularly in Grande, Los Aparejos, Las Peñas, and Negra. Bacillariophyta were positively correlated with higher concentrations of SRP and Si availability. Conversely, *P. chilensis* appeared associated with Copepoda and Cladocera (zooplankton groups), which were also more associated with other algae groups, such as Cyanobacteria and Chlorophyceae. Euglenophyceae and Cryptophyceae were correlated with higher water temperatures, higher conductivity values, and Secchi transparency while Chlorophyceae tend to be restricted to low-conductivity wetlands and low nutrient concentrations. Rotifera was a group poorly represented in the ordination (Figure 8).

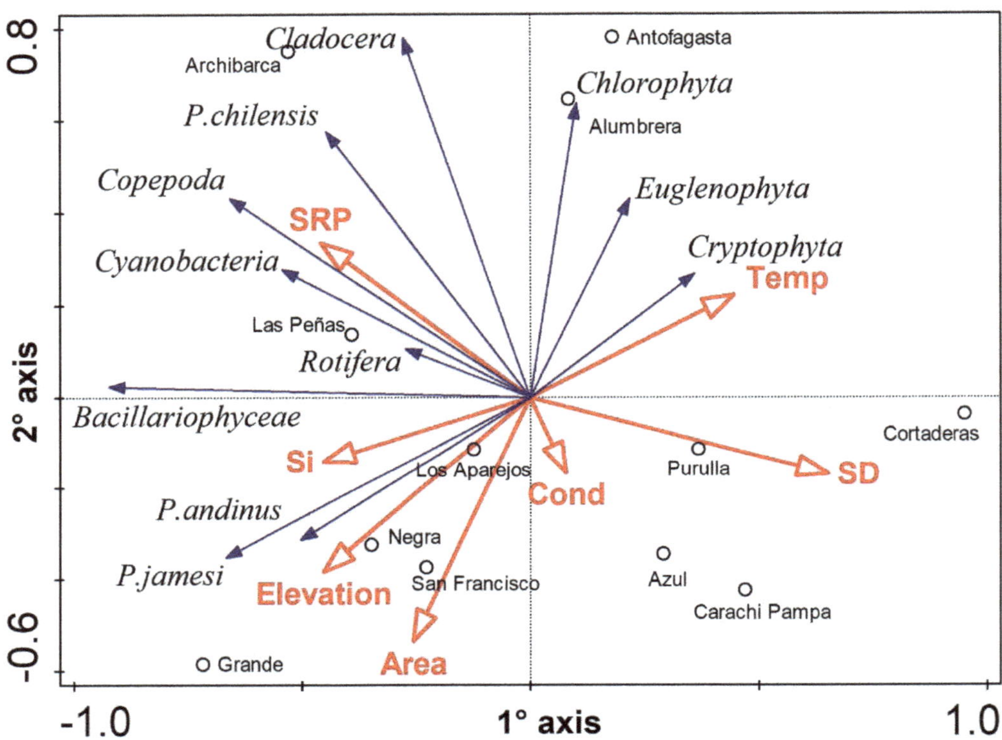

Figure 8. Redundancy analysis (RDA) plot with all the environmental and biological variables included. Abbreviations of variable names as in Table 1.

4. Discussion

Across the Altiplano, most aquatic ecosystems remain poorly described in detail, limiting our ability to understand their ecological importance and sensitivity to human impact. Here, we describe several trophic levels (phytoplankton, zooplankton, and flamingos) along with several limnological and hydroclimatological features for a set of wetlands of a poorly studied region in the High Andean Plateau of Argentina undergoing rapid change from anthropogenic drivers such as climate change and industrial mining development.

4.1. Geological, Hydrological, and Climatological Features

The ecosystems sampled were highly variable in their surface area, the largest being often co-located with salars (evaporite deposits). Most of them (71%) were located at very high elevations, above 3900 m.a.s.l. Owing to the hyperarid climate conditions on many of the floors of these basins, most of the wetlands (67%) were composed of saline or even brine surface water bodies. These conditions are attributable to the rapid uplift of the plateau and progressive orographic-induced aridification, and its location along the Tropic of Capricorn subtropical high belt [88].

The registered nutrient concentrations also appear highly variable among wetlands sampled, with some lagoons such as Archibarca, Cabi, Caro, Hedionda, Peinado, and Las Peñas registering the highest mean values of DIN and SRP. In a similar system of wetlands in the Atacama Desert (Chile), higher nitrate concentrations were attributed to local windblown salts containing nitrate formed by sun UV ray oxidation of nitrogen in the atmosphere or to dust from other nitrate deposits [89]. In some wetlands, such as Archibarca, Cortaderas, and San Francisco, DIN was a limiting nutrient for phytoplankton with values below 100 ug L^{-1}, while SRP always registered values above 10 ug L^{-1}, suggesting it was not limiting [90]. In addition, the DIN/SRP ratio (always < 10) suggested that DIN was the limiting nutrient for those wetlands where data were available during samplings. These results are consistent with those reported for other saline systems around the world e.g., [91–94]. The high variability observed could also be related to fluctuation in hydrological conditions that induce changes in nutrient inputs; therefore, the variations in hydrological conditions are a critical control in semiarid to hyperarid areas, where surface water levels fluctuate seasonally and interannually [95,96]. The wetlands surveyed in our study tend to be very shallow (<1 m in depth), but they experience high variations in water surface area across seasons and years. Decreases in water volume through evaporation result in increases in salinity and nutrient concentrations due to the accumulation of ions [4]. The high values of Si found in most wetlands, and particularly in Grande, Cortaderas, Las Peñas, San Francisco, and Hedionda, are consistent with the tectonically active region and vast exposures of young volcanic rocks [97,98].

The variability in salinity across wetlands is quite dramatic, even between nearby wetlands. Alumbrera and Antofagasta, the freshest of the lagoons sampled, can be explained because both are fed by the Punilla River, a large river through which recent rainwater and snowmelt are efficiently transported toward the basin floor. However, the other wetlands tend to be mainly fed by groundwater recharge and are rich in bicarbonate, sodium, and potassium, which tend to concentrate and precipitate when they reach the surface through springs (here known as vegas and bofedales). This process is fundamental to the creation and maintenance of salars such as Carachi Pampa or Tres Quebradas, where the formation of thick evaporite sequences primarily of halite (NaCl) and dense brine groundwaters require the persistent concentration of salty water close to the surface e.g., [19,99]. For the other wetlands, such as Negra, Grande, or Purulla, variations in salinity of surface waters can be attributed to several processes, including dissolution of old salts, thermal water contributions, evaporative concentration, and mixing between old saline groundwater, and less saline recently recharged or direct meteoric water [89]. For almost all the wetlands sampled, we found high water transparency, with light reaching the bottom of the water body during sampling, especially in Cortaderas and San Francisco. This pattern emphasizes the shallow water column of these ecosystems (<1 m), which contrasts with lakes from areas of the world at similar high elevations, often of glacial origin [100,101].

Our analysis of major ion compositions in these waters provides valuable insight into the distinctions between these wetland systems based on the interactions of inflow waters (meteoric, streamflow, and groundwater), evapoconcentration, and mineral precipitation. The largest group is dominated by Na^+, K^+, and Cl^- reflecting highly geochemically evolved systems. The surface water in these wetlands is sourced from inflow waters that have undergone substantial evapoconcentration along their flow paths and in the water body itself. This group contains two of the saltiest surface waters measured in this work (100 and 230 mS cm^{-1}). The second group is also quite evolved but contains higher proportions of Ca^{++} and Mg^{++}, indicating a distinct geochemical pathway of inflow waters. The other wetlands display some unique geochemical characteristics but can be described primarily as less evolved versions of the other two groups. Alumbrera and Antofagasta, fresh wetlands upgradient of the massive Carchi Pampa basin floor appear to be diluted versions of the waters that feed Carachi Pampa. The primary difference between these systems is that these fresh, upgradient wetlands contain substantial HCO_3^- and $SO4^-$. In these hyperarid to arid evaporite-forming systems, as waters progressively evaporate under intense insolation, first carbonate, then gypsum precipitate leaving very little of these ions in the remaining water [64]. At least from a geochemical perspective, these systems represent two end-member wetland systems but appear to be of highly similar source waters and flow paths. A similar case can be made for Azul, Los Aparejos, and Negra, which reside in adjacent basins, where Los Aparejos represents an intermediate composition but appears to have similar hydrogeological characteristics. This is consistent with basins on the Chilean Altiplano and the Salar de Atacama where dense brines can be explained through the intense and progressive evapoconcentration of much fresher inflow waters near the basin floors either as groundwater exiting the ground from long flow paths or ephemeral flow from precipitation events [21,22,102].

Several important observations can be drawn from our hydroclimate analysis. First, there is a substantial temperature increase at every basin over the last several decades, which is consistent with the effects of global climate change predictions for this region [30]. Increasing temperatures have likely increased overall potential evaporation and transpiration in these wetland ecosystems, perhaps contributing to some of the changes seen in vegetation cover over the same period. Basins with the highest temperature increases also show the strongest declines in vegetation cover, and wetlands with the largest increases in vegetation cover generally show lower temperature increases. Previous studies examining carbon exchange in peatlands show that soil temperature and moisture conditions are well coupled to carbon losses (plant respiration and soil decomposition) and carbon uptake (plant productivity) at an ecosystem level [103,104]. It has also been recently suggested that these ecosystems play an important role in controlling microclimates [11,12]. Though quantifying this impact requires further study, this may be a critical tangible and rapid effect that climate change is having on these wetlands. Changing precipitation patterns have likely come along with increasing temperature. Recent research has shown that this may be currently manifesting in South America by shifting the seasonal monsoon southward, leading to an increase in frequency and magnitude of significant precipitation events on the plateau while also increasing interannual variability [105–107], resulting in increases in frequency and magnitude of large precipitation events but also more pronounced drought periods. Indeed, a major drought has been documented in the region since the mid-2000s and is likely ongoing [108,109]. It is important to incorporate these two climate-driven phenomena into any hydrological or ecological assessment of the region. The changes in basin-scale hydroclimate we describe here can provide a valuable insight into the characteristics of response to natural conditions in each unique system and by extension, current impacts from or potential vulnerability to non-natural disturbances.

The strong decrease in precipitation across all these wetlands is likely primarily due to the ongoing effects of the drought. However, a pronounced regional distinction exists in the magnitude of this decrease, with basins in the southwest showing about double the decrease as those in the northeast. This may be due to the relative proximity of these basins to the dominant moisture source (from the northeast) and/or differences in the magnitude of large events within these basins over the last 5–10 years. The distribution of these large events may also be an important factor in determining the changes in vegetation area in these wetlands. There is increasing evidence of the importance of wetlands as regulators of the local water balance [110,111]. However, we still have little evidence of the effects that these changes in vegetation, through their effects on infiltration and water retention, may have on flamingo distribution patterns and plankton structure (through water level changes, ion concentration, or even modifications in nutrients availability).

Recent research in the region has shown that large precipitation events, which generally occur over only a few days can have a major and lasting influence on wetland surface waters and vegetation extent [112]. In the Tres Quebradas basin, the largest increase in vegetation may also be partially explained by the contribution of increased meltwater from the only glaciers in the region at Monte Pissis and Ojos del Salado [113,114]. These changes in local precipitation and ephemeral flow paths (days to months' time frame) are likely a dominant factor in whether a wetland has grown or shrunk over the last few decades, especially in higher elevation systems with smaller contributing areas.

Another pattern that can be garnered from the distribution of these hydroclimatic changes is the potential importance of the buffering effect that large-scale, old groundwater flow can have on these wetlands. An assessment of the isotope tritium in several of these lagoon waters made by Moran et al. [115] shows that most water flowing through these basins and sustaining these wetlands is old water (at least 70 years old). In this arid environment, fluxes of recent precipitation at the surface or shallow subsurface are spatially rare, but they become focused on specific locations where water tables remain near the surface such as near large rivers such as the Punilla or small perennial streams, *bofadeles* at high elevations, or *vegas* near the margins of wetlands. Although large vegetation changes are observed across the region, most of them can be attributed to the differences in precipitation change from southwest to northeast and the relative amount of the aforementioned old water making its way to these wetlands. Lagoon waters at the floor of the largest and deepest basin (Carachi Pampa) show zero tritium activity, indicating it is sustained nearly entirely by old groundwater. This wetland shows less vegetation cover loss than the wetlands upgradient of it. Since this wetland and others like it (such as Purulla) are likely fed by some of the longest and oldest flow paths in the region, its consistent inflow may buffer this system against the yearly to decadal term climate fluctuation experienced at other wetlands. This concept has important implications for assessing the relative response of individual wetlands in this region to natural change and, therefore, their resilience to potential human impacts such as mining and agricultural activity in the area.

4.2. Biological Communities' Characterization

Flamingo abundance and distribution across wetlands were more highly variable than almost all the variables considered in this study. The three species of flamingos make itinerant use of these wetlands, changing their distribution according to food availability. *Phoenicopterus chilensis* occurs in a wide variety of wetlands, including freshwater and salt lakes, estuaries, and marine coasts from Peru to southern Argentina. It is also distributed from Chile to southern Brazil and Uruguay [55,116,117]. *Phoenicoparrus andinus* and *P. jamesi* are primarily restricted to wetlands in the high Andean plateau of Argentina, Bolivia, Chile, and Peru, although a portion of *P. andinus* population disperses to lowland wetlands in central Argentina during winter, especially when the Andean lakes freeze [53,118]. *Phoenicoparrus jamesi* have been recorded in lowland wetlands in central Argentina [119,120].

Grande is an important wetland for *P. jamesi*, where between 12,000 and 18,000 individuals representing >10% of the global population were recorded. Higher densities of Bacillariophyta (diatoms) were also found in Grande across samples. Grande, Hedionda, and Cortaderas had the highest concentration of Si, a primary element to produce the frustule (cell wall) of diatoms [121,122]. Diatoms also appeared to be controlled by SRP availability, a nutrient that can be rapidly mineralized by flamingos through their feces, constituting a perfect cycle [123]. Indeed, several studies indicate that flamingo distribution is influenced mainly by food abundance and quality e.g., [124,125], which would explain the high numbers of *P. jamesi* in Grande, where the highest density of diatoms has been recorded across years compared with the other wetlands sampled. *P. jamesi* numbers were highly variable in Los Aparejos, Purulla, Las Tunas, and Las Peñas across years, and positively correlated with diatom abundance, which in turn was negatively correlated with the water level of these wetlands. This suggests that increases in water levels in these wetlands negatively affect diatom abundance and therefore food availability.

P. andinus occurred in lower abundance, distributed in several wetlands including Archibarca, Purulla, and San Francisco. *Phoenicopterus chilensis* occurred mainly in Alumbrera, Antofagasta, El Peinado, and Las Peñas, all low to middle-high conductivity wetlands. *Phoenicoparrus jamesi* and *P. andinus* feed mainly on diatoms, while *Phoenicopterus chilensis* has a broader diet including cyanobacteria, insect larvae, and microcrustacean [126–128].

Our analysis shows that both species (*P. andinus* and *P. jamesi*) abundance is linked to diatoms, and *P. chilensis* numbers are linked to large zooplankton (Cladocera and Copepoda) and other algae groups. The evidence does not suggest that flamingos are a controlling factor of plankton abundance in these wetlands, as was concluded in Frau et al. [46]. However, an experiment performed by Hurlbert and Chang [126] suggested that *P. andinus* may influence benthic structure (invertebrates and diatoms), at least locally; therefore, new *in situ* experiments are necessary to elucidate the grazing impact that flamingos may have on the plankton structure in these kinds of wetlands. While the global population of the Chilean Flamingo appears to be stable or increasing [55], *P. chilensis* numbers in the wetlands censused during this study decreased. Therefore, further studies to determine if *P. chilensis* abundance is responding to habitat degradation based on increased human impacts or a consequence of natural cycles in the ecological conditions of these wetlands.

The plankton composition and distribution among the wetlands studies was highly variable. A pattern frequently attributed to salinity and to chemical and temperature gradients that occur within and between these kinds of wetlands [43]. Despite these highly variable patterns registered, the phytoplankton assemblage tended to be dominated mainly by Bacillariophyta (diatoms) as several other studies have reported in similar areas [129–131]. The dominance of genera, such as *Navicula* and *Nitzschia*, both tychoplanktonic algae (adapted to live in stirred waters), is also consistent with several floristics reports [20,36,132,133] in similar environments from the northern part of Argentina. Diatoms were primarily related to silica and SRP availability and middle to high conductivity values. However, some other studies have reported that diatom abundance may be also related to ionic composition [44,134]. Taxonomic affiliations to the species level are difficult. In the Andes region, numerous endemisms comprising a characteristic flora with frequently new species described in several of these reports e.g., [135,136]. For this reason, making this kind of association between ions composition and diatom species could be difficult to reach.

Chlorophyceae was the second most dominant group of algae, frequently represented by just one or two volvaceans species, linked to low nutrient availability and a broad conductivity spectrum, ranging from very low conductance lakes (Antofagasta) to highly saline ones (Tres Quebradas). Less represented groups, such as Euglenophyceae and Cryptophyceae, were generally restricted to high conductivity wetlands and were not in high densities. Their low representation is consistent with what is reported previously in similar environments [137]. It has been speculated that the relative absence of euplanktonic species might be due to the strong UV pressure on these very high lakes [43], with, however, few reports that corroborate this hypothesis e.g., [138]. Remarkably, the presence of euglenoids in saline lagoons has been reported by other authors [40,139,140], indicating that many Euglenophyceae species could be euryhaline or have a broad salinity tolerance.

Cyanobacteria appeared only occasionally and with middle-low densities in Alumbrera, Las Peñas, and San Francisco. In 2013, we recorded a cyanobacteria bloom in Alumbrera dominated by Chroococcales species, which was consistent with potential wastewater infiltration from a nearby plant. No other cyanobacteria blooms were detected since then. Alumbrera has also been increasingly intensively used for livestock (mainly domestic South American camelids), which congregate near the shore of the lake. In recent decades, an increase in urban and agricultural uses of high-elevation wetlands has become more common in the inter-Andean valleys, although the effects of nutrient enrichment on biological communities have not been deeply studied in arid or semiarid zones [95]. Special attention should be given to Alumbrera, a wetland providing services to the town of Antofagasta de la Sierra, to develop sustainable restoration measures and not repeat the eutrophication problems reported by other authors in similar high-elevation ecosystems [141–143].

Our results show that zooplankton was not well represented in species richness or abundance in the Andean plateau wetlands, a pattern that is consistent with the findings of other authors in similar extreme ecosystems, who also conclude that conductivity gradients mainly control zooplankton in these systems [144–147]. Occasionally, one copepod species, *Boechella popoensis* appeared dominant in Archibarca and Las Peñas—a pattern already described by other authors in high elevation saline wetlands from South America [4,143,148]. However, the general tendency was for low-density values of all groups, especially for Rotifers, which were well represented in species but not in abundance. The genera found to be dominant, such as *Boeckella, Cephalodella,* or *Brachionus,* have been reported as saline tolerant in previous studies [145,149,150], which is consistent with the information reported here.

5. Conclusions

This study presents a dataset gathered over ten years and a much broader climatological analysis from 1985 that provides a baseline of the physical-chemical, hydroclimatological, and ecological components and their interactions among a group of high-elevation Andean wetlands. We found that these wetlands are highly variable in salinity (from freshwater to brine), reach high nutrient concentrations, and are very shallow. Our analysis of the geochemical composition of surface waters in these wetlands showed strong distinctions between several groups of wetlands and strong connections across large distances, indicating important regional groundwater connections exist. We also found that these long regional flow paths containing very old water may buffer some wetlands against decadal-term climate variations. Regarding flamingos, *P. jamesi* appears as the dominant species in the study area, primarily because of the high numbers in Grande, and phytoplankton tends to be dominated, in general, by diatoms, while zooplankton is poorly represented. Both plankton groups tend to have low species richness and highly variable abundances among wetlands. The climate is arid, and the general tendency observed in the last decades is increasing air temperatures and decreasing rainfall, with two different trends being observed regarding hydrological responses in the wetlands. One set of wetlands has seen an increase and the other a decrease in the aerial extent of living vegetation. Our data suggest that much of the changes in vegetation patterns are in direct response to the observed climate changes; however, we have no evidence yet of how these changes may affect flamingo and plankton communities. In further studies, we should investigate the effects of these tendencies on the main trophic structure of these wetlands (plankton and flamingos), considering the crucial role that vegetation has in the microclimates of these endorheic basins and, therefore, in the water cycles of these wetlands. In sum, our results suggest that all these wetlands sampled are highly variable and can change rapidly; therefore, further efforts to understand how they function are critical to prevent their degradation and secure the valuable environmental services they provide.

Supplementary Materials: The following are available online at https://www.mdpi.com/article/10.3390/hydrology8040164/s1, Figure S1: Total number of flamingos from the three species recorded reported on each wetland censured from years 2010 to 2020, Table S1: Wetlands sampled and limnological sampling dates, Table S2: Summary of dissolved major ions in surface waters.

Author Contributions: Conceptualization, D.F. and B.J.M.; Methodology, D.F., B.J.M. and P.M.; Formal Analysis, D.F. and B.J.M.; Investigation, D.F., B.J.M., F.A. and P.M.; Resources, P.M., F.A., B.J.M., Y.B., C.M., R.M. and G.M.; Writing—Original Draft Preparation, D.F., B.J.M. and F.A.; Writing—Review & Editing, D.F., B.J.M., P.M., F.A., Y.B., C.M., R.M., G.M. and D.F.B.; Funding Acquisition, P.M., F.A. and B.J.M. All authors have read and agreed to the published version of the manuscript.

Funding: This research received funding from the Disney Wildlife Conservation Fund, BirdLife International, and the US Fish and Wildlife Service.

Acknowledgments: We especially want to thank M. Mosqueira, G. Nuñez, and R. Clark for their valuable help with the fieldwork. As well as Lee Ann Munk and Jordan Jencks for their invaluable assistance in collecting the hydrogeochemical data and Sarah Mcknight and Jordan Jencks for their advice and assistance in refining data extraction from Google Earth Engine.

Conflicts of Interest: The authors declare no conflict of interest.

References

1. Allmendinger, R.W.; Jordan, T.E.; Kay, S.M.; Isacks, B.L. The evolution of the Altiplano-Puna plateau of the Central Andes. *Annu. Rev. Earth Planet. Sci.* **1997**, *25*, 139–174. [CrossRef]
2. Jordan, T.E.; Nester, P.L.; Blanco, N.; Hoke, G.D.; Dávila, F.; TomLinson, A.J. Uplift of the Altiplano-Puna plateau: A view from the west. *Tectonics* **2010**, *29*, TC5007. [CrossRef]
3. Álvarez Blanco, I.; Cejudo-Figueiras, C.; de Godos, I.; Muñoz, R.; Blanco, S. Las diatomeas de los salares del Altiplano boliviano: Singularidades florísticas. *Boletín Real Soc. Española Hist. Nat. Sección Biológica* **2011**, *105*, 67–82.
4. Scott, S.; Dorador, C.; Yanedel, J.P.; Tobar, I.; Hengst, M.; Maya, G.; Harrod, C.; Vila, I. Microbial diversity and trophic components of two high elevation wetlands of the Chilean Altiplano. *Gayana* **2015**, *79*, 45–56. [CrossRef]
5. Williams, W.D.; Boulton, A.J.; Taaffe, R.G. Salinity as a determinant of salt lake fauna: A question of scale. *Hydrobiologia* **1990**, *197*, 257–266. [CrossRef]
6. De la Fuente, A.; Meruane, C.; Suárez, F. Long-term spatiotemporal variability in high Andean wetlands in northern Chile. *Sci. Total Environ.* **2021**, *756*, 143830. [CrossRef]
7. Eugster, H.P. Geochemistry of evaporitic lacustrine deposits. *Annu. Rev. Earth Planet. Sci.* **1980**, *8*, 35–63. [CrossRef]
8. Risacher, F.; Alonso, H.; Salazar, C. The origin of brines and salts in Chilean salars: A hydrochemical review. *Earth-Sci. Rev.* **2003**, *63*, 249–293. [CrossRef]
9. Psenner, R. Alpine waters in the interplay of global change: Complex links-simple effects? In *Global Environmental Change in the Alpine Region. New Horizons in Environmental Economics*; Steininger, K.W., Weck-Hannemann, H., Eds.; Edward Eldgar: Gloucestershire, UK, 2002.
10. Boyle, T.P.; Caziani, S.M.; Waltermire, R.G. Landsat TM inventory and assessment of waterbird habitat in the southern altiplano of South America. *Wetl. Ecol. Manag.* **2004**, *12*, 563–573. [CrossRef]
11. Squeo, F.A.; Warner, B.G.; Aravena, R.; Espinoza, D. Bofedales: High elevatione peatlands of the central Andes. *Rev. Chil. Hist. Nat.* **2006**, *79*, 245–255. [CrossRef]
12. Valois, R.; Schaer, N.; Figueroa, R.; Maldonado, A.; Yáñez, E.; Hevia, A.; Yánez Carrizo, G.; MacDonell, S. Characterizing the Water Storage Capacity and Hydrological Role of Mountain Peatlands in the Arid Andes of North-Central Chile. *Water* **2020**, *12*, 1071. [CrossRef]
13. Gleeson, T.; Marklund, L.; Smith, L.; Manning, A.H. Classifying the water table at regional to continental scales. *Geophys. Res. Lett.* **2011**, *38*, L05401. [CrossRef]
14. Walvoord, M.A.; Plummer, M.A.; Phillips, F.M.; Wolfsberg, A.V. Deep arid system hydrodynamics: 1. Equilibrium states and response times in thick desert vadose zones. *Water Resour. Res.* **2002**, *38*, 1308. [CrossRef]
15. Liu, Y.; Wagener, T.; Beck, H.E.; Hartmann, A. What is the hydrologically effective area of a catchment? *Environ. Res. Lett.* **2020**, *15*, 104024. [CrossRef]
16. Boutt, D.F.; Corenthal, L.G.; Moran, B.J.; Munk, L.A.; Hynek, S.A. Imbalance in the modern hydrologic budget of topographic catchments along the western slope of the Andes (21–25°S): Implications for groundwater recharge assessment. *Hydrogeol. J.* **2021**, *29*, 985–1007. [CrossRef]
17. Alley, W.M.; Healy, R.W.; LaBaugh, J.W.; Reilly, T.E. Flow and storage in groundwater systems. *Science* **2002**, *296*, 1985–1990. [CrossRef]
18. Maxwell, R.M.; Condon, L.E.; Kollet, S.J.; Maher, K.; Haggerty, R.; Forrester, M.M. The imprint of climate and geology on the residence times of groundwater. *Geophys. Res. Lett.* **2016**, *43*, 701–708. [CrossRef]
19. Corenthal, L.G.; Boutt, D.F.; Hynek, S.A.; Munk, L.A. Regional groundwater flow and accumulation of a massive evaporite deposit at the margin of the Chilean Altiplano. *Geophys. Res. Lett.* **2016**, *43*, 8017–8025. [CrossRef]
20. Munk, L.A.; Boutt, D.F.; Hynek, S.A.; Moran, B.J. Hydrogeochemical fluxes and processes contributing to the formation of lithium-enriched brines in a hyper-arid continental basin. *Chem. Geol.* **2018**, *493*, 37–57. [CrossRef]
21. Moran, B.J.; Boutt, D.F.; Munk, L.A. Stable and Radioisotope Systematics Reveal Fossil Water as Fundamental Characteristic of Arid Orogenic-Scale Groundwater Systems. *Water Resour. Res.* **2019**, *55*, 11295–11315. [CrossRef]
22. Munk, L.A.; Boutt, D.F.; Moran, B.J.; McKnight, S.V.; Jenckes, J. Hydrogeologic and Geochemical Distinctions in Freshwater-Brine Systems of an Andean Salar. *Geochem. Geophys.* **2021**, *22*, e2020GC009345.
23. Izquierdo, A.E.; Aragón, R.; Navarro, C.J.; Casagranda, E. Humedales de la Puna: Principales proveedores de servicios ecosistémicos de la región. In *Serie de Conservación de la Naturaleza 24: La Puna Argentina: Naturaleza y Cultura*; Grau, H.R., Babot, M.J., Izquierdo, A., Grau, A., Eds.; Fundación Miguel Lillo: Tucumán, Argentina, 2018; pp. 96–111.
24. Gluzman, G. Minería y metalurgia en la antigua gobernación del Tucumán (siglos XVI–XVII): Colonial Tucumán 16th and 17th Centuries. *Mem. Am.* **2007**, *15*, 157–184.
25. Kesler, S.E.; Gruber, P.W.; Medina, P.A.; Keoleian, G.A.; Everson, M.P.; Wallington, T.J. Global lithium resources: Relative importance of pegmatite, brine and other deposits. *Ore Geol. Rev.* **2012**, *48*, 55–69. [CrossRef]
26. Munk, L.A.; Hynek, S.A.; Bradley, D.; Boutt, D.F.; Labay, K.; Jochens, H. Lithium brines: A global perspective. *Rev. Econ. Geol.* **2016**, *18*, 339–365.
27. Williams, W.D. Environmental threats to salt lakes and the likely status of inland saline ecosystems in 2025. *Environ. Conserv.* **2002**, *29*, 154–167. [CrossRef]

28. Tolotti, M.; Thies, M.; Cantonati, M.; Hansen, C.; Thaler, B. Flagellate algae (Chrysophyceae, Dinophyceae, Cryptophyceae) in 48 high mountain lakes of the Northern and Southern slope of the Eastern Alps: Biodiversity, taxa distribution and their driving variables. *Hydrobiologia* **2006**, *502*, 331–348. [CrossRef]

29. Meehl, G.A.; Arblaster, J.M.; Tebaldi, C. Understanding future patterns of increased precipitation intensity in climate model simulations. *Geophys. Res. Lett.* **2005**, *32*, L18719. [CrossRef]

30. Urrutia, R.; Vuille, M. Climate change projections for the tropical Andes using a regional climate model: Temperature and precipitation simulations for the end of the 21st century. *J. Geophys. Res.* **2009**, *114*, D02108. [CrossRef]

31. Barros, V.R.; Boninsegna, J.A.; Camilloni, I.A.; Chidiak, M.; Magrín, G.O.; Rusticucci, M. Climate change in Argentina: Trends, projections, impacts and adaptation. *Wiley Interdiscip. Rev. Clim. Chang.* **2015**, *6*, 151–169. [CrossRef]

32. Pabón-Caicedo, J.D.; Arias, P.A.; Carril, A.F.; Espinoza, J.C.; Borrel, L.F.; Goubanova, K.; Lavado-Casimiro, W.; Masiokas, M.; Solman, S.; Villalba, R. Observed and Projected Hydroclimate Changes in the Andes. *Front. Earth Sci.* **2020**, *8*, 61. [CrossRef]

33. Ibarguchi, G. From Southern Cone arid lands, across Atacama, to the Altiplano: Biodiversity and conservation at the ends of the world. *Biodiversity* **2014**, *15*, 255–264. [CrossRef]

34. Aguilar, P.; Dorador, C.; Vila, I.; Sommaruga, R. Bacterioplankton composition in tropical high-elevation lakes of the Andean plateau. *FEMS Microbiol. Ecol.* **2018**, *94*, fiy004. [CrossRef]

35. Farías, E. *Microbial Ecosystems in Central Andes Extreme Environments*; Springer International Publishing: Cham, Switzerland, 2020.

36. Locascio de Mitrovic, C.A.; Villagra de Gamundi, A.; Juárez, J.; Ceraolo, M. Características limnológicas y zooplancton de cinco lagunas de la Puna-Argentina. *Ecol. Boliv.* **2005**, *40*, 12–26.

37. Aguilera, X.; Declercka, S.; De Meester, L.; Maldonado, M.; Ollevier, F. Tropical high Andes lakes: A limnological survey and an assessment of exotic rainbow trout (*Oncorhynchus mykiss*). *Limnologica* **2006**, *36*, 258–268. [CrossRef]

38. Seeligmann, C.; Maidana, N.I.; Morales, M. Diatomeas (Bacillariophyceae) de humedales de altura de la provincia de Jujuy-Argentina. *Boletín Soc. Argent. Botánica* **2008**, *43*, 1–17.

39. Maidana, N.I.; Seeligmann, C.T.; Morales, M.R. Bacillariophyceae del Complejo Lagunar Vilama (Jujuy, Argentina). *Boletín Soc. Argent. Botánica* **2009**, *44*, 257–271.

40. Mirande, V.; Tracanna, B. Estructura y controles abióticos del fitoplancton en humedales de altura. *Ecol. Austral* **2009**, *19*, 119–128.

41. Declerck, S.A.J.; Coronel, J.S.; Legendre, P.; Brendonck, L. Scale dependency of processes structuring metacommunities of cladocerans in temporary pools of High-Andes wetlands. *Ecography* **2011**, *34*, 296–305. [CrossRef]

42. Caputo, L.; Giuseppe, A.; Givovich, A. Limnological features of Laguna Teno (35° S, Chile): A high-elevation lake impacted by volcanic activity. *Fundam. Appl. Limnol.* **2013**, *183*, 323–335. [CrossRef]

43. Albarracín, V.H.; Kurth, D.; Ordoñez, O.F.; Belfiore, C.; Luccini, E.; Salum, G.M.; Piacentini, R.D.; Farías, M.E. High-Up: A Remote Reservoir of Microbial Extremophiles in Central Andean Wetlands. *Front. Microbiol.* **2015**, *6*, 1404. [CrossRef]

44. Servant-Vildary, S.; Roux, M. Multivariate analysis of diatoms and water chemistry in Bolivian saline lakes. In *Saline Lakes Developments in Hydrobiology*; Comínand, F., Northcote, T., Eds.; Springer: Leiden, The Netherlands, 1990; pp. 267–290.

45. Rejas, D.; Valverde, C.; Fernández, C.E. Limitación por nutrientes y pastoreo como factores de control de las densidades de bacterias y algas planctónicas en una laguna altoandina (Cochabamba, Bolivia). *Rev. Boliv. Ecol. Conserv. Ambient.* **2012**, *30*, 1–12.

46. Frau, D.; Battauz, Y.; Mayora, G.; Marconi, P. Controlling factors in planktonic communities over a salinity gradient in high-elevation lakes. *Ann. Limnol.-Int. J. Limnol.* **2015**, *51*, 261–272.

47. Fjeldså, J.; Krabbe, N.K. *Birds of the High Andes*; Museum Tusculanum Press: Copenhagen, Denmark, 1990; 881p.

48. Caziani, S.M.; Derlindati, E.J.; Tálamo, A.; Sureda, A.L.; Trucco, C.E.; Nicolossi, G. Waterbird richness in altiplano wetlands of northwestern Argentina. *Waterbirds* **2001**, *24*, 103–117. [CrossRef]

49. Vuilleumier, F.; Simberloff, D. Ecology versus history as determinants of patchy and insular distribution in high Andean birds. In *Evolutionary Biology*; Hecht, M., Steere, W., Wallace, B., Eds.; Plenum Publishing Corporation: New York, NY, USA, 1980; pp. 235–379.

50. Tellería, J.L.; Venero, J.L.; Santos, T. Conserving birdlife of Peruvian highland bogs: Effects of patch-size and habitat quality on species richness and bird numbers. *Ardeola* **2006**, *53*, 271–283.

51. Jacobsen, D.; Dangles, O. *Ecology of High. Elevation Waters*; Oxford University Press: Oxford, MS, USA, 2017.

52. Castellino, M.; Lesterhuis, A. *Censo Simultáneo de Falaropos 2020-Resumen y Resultados*; Manomet-WHSRN: Plymouth, MA, USA, 2020.

53. Caziani, S.M.; Rocha, O.; Rodríguez Ramírez, E.; Romano, M.C.; Derlindati, E.J.; Tálamo, A.; Ricalde, D.; Quiroga, C.; Contreras, J.P.; Valqui, M.; et al. Seasonal distribution, abundance, and nesting of Puna, Andean, and Chilean flamingos. *Condor* **2007**, *109*, 276–287. [CrossRef]

54. Marconi, P.; Sureda, A.L.; Arengo, F.; Aguilar, M.S.; Amado, N.; Alza, L.; Rocha, O.; Torres, R.; Moschione, F.; Romano, M.; et al. Fourth simultaneous flamingo census in South America: Preliminary results. *Flamingo* **2011**, *18*, 48–53.

55. Marconi, P.; Arengo, F.; Castro, A.; Rocha, O.; Valqui, M.; Aguilar, S.; Barberis, I.; Castellino, M.; Castro, L.; Derlindati, E.; et al. Sixth International Simultaneous Census of three flamingo species in the Southern Cone of South America: Preliminary Analysis. *Flamingo* **2020**, *12*, 67–75.

56. Villagrán, C.; Castro, V. *Ciencia Indígena de Los Andes del Norte de Chile*; Editorial Universitaria: Santiago, Chile, 1997; p. 362.

57. Catalan, J.; Rondón, J.C.D. Perspectives for an integrated understanding of tropical and temperate high-mountain lakes. *J. Limnol.* **2016**, *75*, 215–234. [CrossRef]

58. Adrian, R.; O'Reilly, C.M.; Zagarese, H.; Baines, S.B.; Hessen, D.O.; Keller, W.; Livingstone, D.M.; Sommaruga, R.; Straile, D.; Van Donk, E.; et al. Lakes as sentinels of climate change. *Limnol. Oceanogr.* **2009**, *54*, 2283–2297. [CrossRef]
59. Coronel, J.S.; Declerck, S.; Maldonado, M.; Ollevier, F.; Brendonck, L. Temporary shallow pools in the high-Andes 'bofedal' peat lands: A limnological characterization at different spatial scales. *Arch. Sci.* **2004**, *57*, 85–96.
60. Wanger, T.C. The Lithium future—Resources, recycling, and the environment. *Conserv. Lett.* **2011**, *4*, 202–206. [CrossRef]
61. Marconi, P.; Arengo, F.; Clark, A. The arid Andean plateau waterscapes and the lithium triangle: Flamingos as flagships for conservation of wetlands under pressure from mining development. *Wetl. Ecol. Manag.* **2021**, under review.
62. Morlans, M.C. Regiones Naturales de Catamarca. Provincias Geológicas y Provincias Fitogeográficas. *Rev. Cienc. Técnica* **1995**, *2*, 1–42.
63. Pekel, J.F.; Cottam, A.; Gorelick, N.; Belward, A.S. High-resolution mapping of global surface water and its long-term changes. *Nature* **2016**, *540*, 418–422. [CrossRef]
64. Warren, J.K. *Evaporites: A Geological Compendium*; Springer International Publishing: Cham, Switzerland, 2016; p. 1813.
65. Hilton, J.; Rigg, E. Determination of nitrate in lake water by the adaptation of the hydrazine-copper reduction method for use on a discrete analyser: Performance statistics and an instrument-induced difference from segmented flow conditions. *Analyst* **1993**, *108*, 1026–1028. [CrossRef]
66. APHA. *Standard Methods for the Examination of Water and Wastewater*; American Public Health Association: Washington, DC, USA, 2005.
67. Sureda, A.L. Patrones de Diversidad en Aves de Lagunas Altoandinas de Catamarca, Noroeste de Argentina. Licentiate Thesis, Universidad Nacional de Salta, Salta, Argentina, 2003; p. 63.
68. Lehner, B.; Grill, G. Global River hydrography and network routing: Baseline data and new approaches to study the world's large river systems. *Hydrol. Process.* **2013**, *27*, 2171–2186. [CrossRef]
69. Abatzoglou, J.T.; Dobrowski, S.Z.; Parks, S.A.; Hegewisch, K.C. TerraClimate, a high-resolution global dataset of monthly climate and climatic water balance from 1958–2015. *Sci. Data* **2018**, *5*, 170191. [CrossRef]
70. McNally, A.; Arsenault, K.; Kumar, S.; Shukla, S.; Peterson, P.; Wang, S.; Funk, C.; Peters-Lidard, C.D.; Verdin, J.P. A land data assimilation system for sub-Saharan Africa food and water security applications. *Sci. Data* **2017**, *4*, 170012. [CrossRef]
71. Tucker, C.J. Red and photographic infrared linear combinations for monitoring vegetation. *Remote Sens. Environ.* **1979**, *8*, 127–150. [CrossRef]
72. Utermöhl, H. Zur Vervollkommnung der quantitativen Phytoplankton. *Methodik. Mitt. Int. Verein. Limnol.* **1958**, *9*, 1–38.
73. Venrick, E.L. How many cells to count? In *Phytoplankton Manual*; Von Sournia, A., Ed.; UNSECO: Paris, France, 1978; pp. 167–180.
74. Lee, R.D. *Phycology*; Cambridge University Press: Cambridge, UK, 2008.
75. Krammer, K.; Lange-Bertalot, H. Bacillariophyceae. 3. Teil Centrales, Fragilariaceae, Eunotiaceae. In *Süsswasserflora von Mitteleuropa*; Ettl, H., Gerloff, J., Heynig, H., Mollenhauer, D., Eds.; Gustav Fischer Verlag: Stuttgar, Germany, 1991.
76. Tell, G.; Conforti, V. Euglenophyta pigmentadas da Argentina. *Bibl. Phycol.* **1986**, *75*, 1–301.
77. Komárek, J.; Fott, B. Chlorophyceae, chlorococcales. In *Das Phytoplankton des Sdwasswes. Die Binnenggewasser*; Huber-Pestalozzi, G., Ed.; Schweizerbart'sche Verlagsbuchhandlung: Stuttgart, Germany, 1983.
78. Komárek, J.; Anagnostidis, K. Cyanoprokariota. 1. Chroococcales. In *Subwasserflora von Mitteleuropa*; Ett, H., Gärdner, G., Heynig, H., Mollenhauer, D., Eds.; Gustav Fischer Verlag: Stuttgar, Germany, 1999.
79. Komárek, J.; Anagnostidis, K. Cyanoprokaryota. Teil 2: Oscillatoriales. In *Süsswasserflora von Mitteleuropa 19/2*; Büdel, B., Gärtner, G., Krienitz, L., Schagerl, M., Eds.; Elsevier: München, Germany, 2005.
80. Ahlstrom, E.H. A revision of the Rotatorian genera *Brachionus* and *Platyias* with descriptions of one new species and two new varieties. *Bull. Am. Mus. Nat. Hist.* **1940**, *77*, 143–148.
81. Koste, W. *Rotatoria. Die Radertiere Mitteleuropas*; Gebruder Borntraeger: Berlin, Germany, 1978.
82. Kořínek, V. *Diaphanosoma birgei n.sp.* (Crustacea, Cladocera). A new species from America and its widely distributed subspecies *Diaphanosoma birgei* ssp. *lacustris* n.ssp. *Can. J. Zool.* **1981**, *59*, 1115–1121. [CrossRef]
83. Kořínek, V. Cladocera. In *A Guide to Tropical Freshwater Zooplankton*; Fernando, C.H., Ed.; Backhuys Publishers: Leiden, The Netherlands, 2002; pp. 69–122.
84. Korovchinski, N.M. Sididae and Holopedidae (Crustacea: Daphniiformes). In *Guides to Identification of the Microinvertebrates of the Continental Waters of the World*; SPB Academic Publishers: The Hague, The Netherlands, 1992.
85. Alekseev, V.R. Copepoda. In *A Guide to Tropical. Freshwater Zooplankton*; Fernando, C.H., Ed.; Backhugs Publishers: Leiden, The Netherlands, 2002; pp. 123–188.
86. Marconi, P. Técnicas de monitoreo de condiciones ecológicas en la Red de Humedales de Importancia para la Conservación de Flamencos Altoandinos. In *Manual de Técnicas de Monitoreo de Condiciones Ecológicas para el Manejo Integrado de la red de Humedales de Importancia para la Conservación de Flamencos Altoandinos*; Marconi, P., Ed.; Fundación YUCHAN: Salta, Argentina, 2010; pp. 8–14.
87. Messager, M.L.; Lehner, B.; Grill, G.; Nedeva, I.; Schmitt, O. Estimating the volume and age of water stored in global lakes using a geo-statistical approach. *Nat. Commun.* **2016**, *7*, 13603. [CrossRef] [PubMed]
88. Pingel, H.; Alonso, R.N.; Altenberger, U.; Cottle, J.; Strecker, M.R. Miocene to Quaternary basin evolution at the southeastern Andean Plateau (Puna) margin (ca. 24°S lat, Northwestern Argentina). *Basin Res.* **2019**, *31*, 808–826. [CrossRef]
89. Gamboa, C.; Godfrey, L.; Herrera, C.; Custodio, E.; Soler, A. The origin of solutes in groundwater in a hyper-arid environment: A chemical and multi-isotope approach in the Atacama Desert, Chile. *Sci. Total Environ.* **2019**, *690*, 329–351. [CrossRef]

90. Reynolds, C.S. *The Ecology of Phytoplankton*; Cambridge University Press: Cambridge, UK, 2006; p. 552.
91. Cloern, J.E.; Alpine, A.E.; Cole, B.E.; Wong, R.L.J.; Arthur, J.F.; Ball, M.D. River discharge controls phytoplankton dynamics in the northern San Francisco Bay estuary. *Estuar. Coast. Shelf Sci.* **1983**, *16*, 415–429. [CrossRef]
92. Herbst, D.B.; Bradley, T.J. A Malpighian tubule lime gland in an insect inhabiting alkaline salt lakes. *J. Exp. Biol.* **1989**, *145*, 63–78. [CrossRef]
93. Salm, C.R.; Saros, J.E.; Martin, C.S.; Erickson, J.M. Patterns of seasonal phytoplankton distribution in prairie saline lakes of the northern Great Plains (U.S.A.). *Saline Syst.* **2009**, *5*, 1. [CrossRef] [PubMed]
94. Salm, C.R.; Saros, J.E.; Fritz, S.C.; Osburn, C.L.; Reineke, D.M. Phytoplankton productivity across prairie saline lakes of the Great Plains (USA): A step toward deciphering patterns through lake classification models. *Can. J. Fish. Aquat. Sci.* **2009**, *66*, 1435–1448. [CrossRef]
95. Sánchez Carrillo, S.; Álvarez-Cobelas, M. Nutrient Dynamics and Eutrophication Patterns in A Semi-Arid Wetland: The Effects of Fluctuating Hydrology. *Water Air Soil Pollut.* **2001**, *131*, 97–118. [CrossRef]
96. Sánchez Carrillo, S.; Angeler, D.G.; Alvarez-Cabelas, M.; Sanchez-Andres, R. Freshwater wetland eutrophication. In *Eutrophication: Causes, Consequences and Control*; Ansari, A.A., Singh Gill, S., Lanza, G.R., Rast, W., Eds.; Springer: Heidelberg, Germany, 2011; pp. 195–210.
97. Jordan, T.E.; Mpodozis, C.; Muñoz, N.; Blanco, N.; Pananont, P.; Gardeweg, M. Cenozoic subsurface stratigraphy and structure of the Salar de Atacama Basin, northern Chile. *J. S. Am. Earth Sci.* **2007**, *23*, 122–146. [CrossRef]
98. Salisbury, M.J.; Jicha, B.R.; de Silva, S.L.; Singer, B.S.; Jiménez, N.C.; Ort, M.H. 40Ar/39Ar chronostratigraphy of Altiplano-Puna volcanic complex ignimbrites reveals the development of a major magmatic province. *Geol. Soc. Am. Bull.* **2011**, *123*, 821–840. [CrossRef]
99. Finstad, K.; Pfeiffer, M.; McNicol, G.; Barnes, J.; Demergasso, C.; Chong, G.; Amundson, R. Rates and geochemical processes of soil and salt crust formation in Salars of the Atacama Desert, Chile. *Geoderma* **2016**, *284*, 57–72. [CrossRef]
100. Leopold, L.B. Temperature profiles and bathymetry of some high mountain lakes. *Proc. Natl. Acad. Sci. USA* **2000**, *97*, 6267–6270. [CrossRef] [PubMed]
101. Muñoz, R.; Huggel, C.; Frey, H.; Cochachin, A.; Haeberli, W. Glacial Lake depth and volume estimation based on a large bathymetric dataset from the Cordillera Blanca, Peru. *Earth Surf. Process. Landf.* **2020**, *45*, 1510–1527. [CrossRef]
102. Boutt, D.F.; Hynek, S.A.; Munk, L.A.; Corenthal, L.G. Rapid recharge of fresh water to the halite-hosted brine aquifer of Salar de Atacama, Chile. *Hydrol. Process.* **2016**, *30*, 4720–4740. [CrossRef]
103. Price, J.S. Hydrology and microclimate of a partly restored cutover bog, Quebec. *Hydrol. Process.* **1996**, *10*, 1263–1272. [CrossRef]
104. Plach, J.M.; Petrone, R.M.; Waddington, J.M.; Kettridge, N.; Devito, K.J. Hydroclimatic influences on peatland CO_2 exchange following upland forest harvesting on the Boreal Plains. *Ecohydrology* **2016**, *9*, 1590–1603. [CrossRef]
105. Jordan, T.E.; Herrera, L.C.; Godfrey, L.V.; Colucci, S.J.; Gamboa, P.C.; Urrutia, M.J.; González, G.L.; Paul, J.F. Isotopic characteristics and paleoclimate implications of the extreme precipitation event of March 2015 in Northern Chile. *Andean Geol.* **2019**, *46*, 1–31. [CrossRef]
106. Langenbrunner, B.; Pritchard, M.S.; Kooperman, G.J.; Randerson, J.T. Why does Amazon precipitation decrease when tropical forests respond to increasing CO_2? *Earth's Future* **2019**, *7*, 450–468. [CrossRef]
107. Pascale, S.; Carvalho, L.M.V.; Adams, D.K.; Castro, C.L.; Cavalcanti, I.F.A. Current and future variations of the monsoons of the Americas in a warming climate. *Curr. Clim. Chang. Rep.* **2019**, *5*, 125–144. [CrossRef]
108. Garreaud, R.D.; Boisier, J.P.; Rondanelli, R.; Montecinos, A.; Sepúlveda, H.H.; Veloso-Aguila, D. The Central Chile Mega Drought (2010–2018): A climate dynamics perspective. *Int. J. Climatol.* **2020**, *40*, 421–439. [CrossRef]
109. Ferrero, M.E.; Villalba, R. Interannual and long-term precipitation variability along the subtropical mountains and adjacent Chaco (22–29° S) in Argentina. *Front. Earth Sci.* **2019**, *7*, 148. [CrossRef]
110. Petrone, R.M.; Devito, K.J.; Silins, U.; Mendoza, C.; Brown, S.C.; Kaufman, S.C.; Price, J.S. Transient peat properties in two pond-peatland complexes in the sub-humid Western Boreal Plain, Canada. *Mires Peat* **2008**, *3*, 1–13.
111. Rolando, J.L.; Turina, C.; Ramírez, D.A.; Maresa, V.; Monerris, J.; Quiroz, R. Key ecosystem services and ecological intensification of agriculture in the tropical high-Andean Puna as affected by land-use and climate changes. *Agric. Ecosyst. Environ.* **2017**, *236*, 221–233. [CrossRef]
112. Moran, B.J.; Boutt, D.F.; McKnight, S.V.; Jenckes, J.; Munk, L.A.; Corkran, D.; Kirshen, A. *Water Sustainability, Drought, Relic Groundwater and Lithium Resource Extraction in an Arid Landscape*; University of Massachusetts: Amherst, MA, USA, 2021; Manuscript to be submitted.
113. García, A.; Ulloa, C.; Amigo, G.; Milana, J.P.; Medina, C. An inventory of cryospheric landforms in the arid diagonal of South America (high Central Andes, Atacama region, Chile). *Quat. Int.* **2017**, *438*, 4–19. [CrossRef]
114. Schaffer, N.; MacDonell, S.; Réveillet, M.; Yáñez, E.; Valois, R. Rock glaciers as a water resource in a changing climate in the semiarid Chilean Andes. *Reg. Environ. Chang.* **2019**, *19*, 1263–1279. [CrossRef]
115. Moran, B.J.; Boutt, D.F.; Munk, L.A.; Fisher, J.D. Pronounced Water Age Partitioning Between Arid Andean Aquifers and Fresh-Saline Lagoon Systems. In Proceedings of the EGU General Assembly 2021a, online/virtual, 19–30 April 2021. EGU21-13753. [CrossRef]
116. Canevari, P. *El Flamenco Común*; Centro Editor de América Latina: Buenos Aires, Argentina, 1983.

117. Bucher, E.H. Flamencos. In *Bañados del río Dulce y Laguna Mar Chiquita*; Bucher, E.H., Ed.; Academia Nacional de Ciencias: Córdoba, Argentina, 2006; pp. 251–261.
118. Romano, M.; Pagano, F.; Luppi, M. Registros de Parina grande (*Phoenicopterus andinus*) en la laguna Melincué, Santa Fe, Argentina. *Nuestras Aves* **2002**, *43*, 15–17.
119. Cruz, N.; Barisón, C.; Romano, M.; Arengo, F.; Derlindati, E.; Barberis, I. A new record of James's flamingo (*Phoenicoparrus jamesi*) from Laguna Melincué, a lowland wetland in East-Central Argentina. *Wilson J. Ornithol.* **2013**, *125*, 217–221. [CrossRef]
120. Torres, R.M.; Michelutti, P. Aves Acuáticas. In *Bañados del río Dulce y Laguna Mar Chiquita (Córdoba, Argentina)*; Bucher, E.H., Ed.; Academia Nacional de Ciencias: Córdoba: Argentina, 2006; pp. 237–249.
121. Garnier, J.; Beusen, A.; Thieu, V.; Billen, G.; Bouwman, L. N:P:Si nutrient export ratios and ecological consequences in coastal seas evaluated by the ICEP approach. *Glob. Biogeochem. Cycles* **2010**, *24*, GB0A05. [CrossRef]
122. Tavernini, S.; Pierobon, E.; Viaroli, P. Physical factors and dissolved reactive silica affect phytoplankton community structure and dynamics in a lowland eutrophic river (Po River, Italy). *Hydrobiologia* **2011**, *669*, 213–225. [CrossRef]
123. Laguna, C.; López-Perea, J.J.; Feliu, J.; Jiménez-Moreno, M.; Rodríguez Martín-Doimeadios, R.C.; Florín, M.; Mateo, R. Nutrient enrichment and trace element accumulation in sediments caused by waterbird colonies at a Mediterranean semiarid floodplain. *Sci. Total Environ.* **2021**, *777*, 145748. [CrossRef]
124. Henriksen, M.V.; Hangstrup, S.; Work, F.; Krogsgaard, M.K.; Groom, G.B.; Fox, A.D. Flock distributions of Lesser Flamingos Phoeniconaias minor as potential responses to food abundance-predation risk trade-offs at Kamfers Dam, South Africa. *Wildfowl* **2015**, *65*, 3–18.
125. Krienitz, L.; Krienitz, D.; Dadheech, P.K.; Hübener, T.; Kotut, K.; Luo, W.; Teubner, R.; Versfeld, W.D. Food algae for Lesser Flamingos: A stocktaking. *Hydrobiologia* **2016**, *775*, 2–50. [CrossRef]
126. Hurlbert, S.H.; Chang, C. Ornitholimnology: Effects of grazing by the Andean Flamingo (Phoenicoparrus andinus). *Proc. Natl. Acad. Sci. USA* **1983**, *80*, 4766–4769. [CrossRef] [PubMed]
127. Mascitti, V.; Kravetz, F.O. Bill morphology of South American flamingos. *Condor* **2002**, *104*, 73–83. [CrossRef]
128. Tobar, C.N.; Rau, J.R.; Fuentes, N.; Gantz, A.; Suazo, C.G.; Cursach, J.A.; Santibañez, A.; Pérez Schultheiss, J. Diet of the Chilean flamingo *Phoenicopterus chilensis* (Phoenicopteriformes: Phoenicopteridae) in a coastal wetland in Chiloé, southern Chile. *Rev. Chil. Hist. Nat.* **2014**, *87*, 1–7. [CrossRef]
129. Sylvestre, F.; Servant-Vildary, S.; Roux, M. Diatom-based ionic concentration and salinity models from the south Bolivian Altiplano (15-23°S). *J. Paleolimnol.* **2001**, *25*, 279–295. [CrossRef]
130. Tapia, P.M.; Fritz, S.C.; Baker, P.A.; Seltzer, G.O.; Dunbar, R.B. A Late Quaternary diatom record of tropical climatic history from Lake Titicaca (Peru and Bolivia). *Palaeogeogr. Palaeoclimatol. Palaeoecol.* **2003**, *194*, 139–164. [CrossRef]
131. Tapia, P.M.; Fritz, S.C.; Seltzer, G.O.; Rodbell, D.T.; Metivier, S.P. Contemporary distribution and late-quaternary stratigraphy of diatoms in the Junin plain, central Andes, Peru. *Boletín Soc. Geológica Peru* **2006**, *101*, 19–42.
132. Maidana, N.; Seeligmann, C. Diatomeas (Bacillariophyceae) de ambientes acuáticos de altura de la Provincia de Catamarca, Argentina, II. *Boletín Soc. Argent. Botánica* **2006**, *41*, 1–13.
133. Maidana, N.I.; Seeligmann, C.; Morales, M.R. El género Navicula *sensu stricto* (Bacillariophyceae) en humedales de altura de Jujuy, Argentina. *Boletín Soc. Argent. Botánica* **2011**, *46*, 13–29.
134. Saros, J.E.; Fritz, S.E. Changes in the growth rates of saline-lake diatoms in response to variation in salinity, brine type, and nitrogen form. *J. Plankton Res.* **2000**, *22*, 1071–1083. [CrossRef]
135. Díaz, C.A.; Maidana, N.I. A new monoraphid diatom genus: *Haloroundia* Diaz and Maidana. *Nova Hedwig. Beih.* **2006**, *130*, 177–183.
136. Blanco, S.; Álvarez-Blanco, I.; Cejudo-Figueiras, C.; DeGodos, I.; Bécares, E.; Muñoz, R.; Guzmán Suarez, H.; Vargas, V.; Soto, R. New diatom taxa from high-altitude Andean saline lakes. *Diatom Res.* **2013**, *28*, 13–27. [CrossRef]
137. Cabrol, N.A.; Grin, E.A.; Chong, G.; Minkley, E.; Hock, A.N.; Yu, Y.; Bebout, L.; Fleming, E.; Häder, D.P.; Demergasso, C.; et al. The high-lakes project. *J. Geophys. Res.* **2007**, *114*, G00D06. [CrossRef]
138. Bastidas Navarro, M.; Balseiro, E.; Modenutti, B. UV radiation simultaneously affects phototrophy and phagotrophy in nanoflagellate-dominated phytoplankton from an Andean shallow lake. *Photochem. Photobiol. Sci.* **2011**, *10*, 1318. [CrossRef]
139. Hammer, U.T.; Shamess, J.; Haynes, R.C. The distribution and abundance of algae in saline lakes of Saskatchewan, Canada. *Hydrobiologia* **1983**, *105*, 1–26. [CrossRef]
140. Padisák, J.; Dokulil, M. Meroplankton dynamic in a saline, turbulent, turbid shallow lake (Neusiedlersee, Austria and Hungary). *Hydrobiologia* **1994**, *289*, 23–42. [CrossRef]
141. Zúñiga, L.R.; Campos, V.; Pinochet, H.; Prado, B. A limnological reconnaissance of Lake Tebenquiche, Salar de Atacama, Chile. *Hydrobiologia* **1991**, *210*, 19–24. [CrossRef]
142. Gunkel, G. Limnology of an Equatorial High Mountain Lake in Ecuador, Lago San Pablo. *Limnologica* **2000**, *30*, 113–120. [CrossRef]
143. Van Colen, W.; Portilla, K.; Oñab, T.; Wyseurec, G.; Goethalsd, P.; Velarde, E.; Muylaerta, K. Limnology of the neotropical high elevation shallow lake Yahuarcocha (Ecuador) and challenges for managing eutrophication using biomanipulation. *Limnologica* **2017**, *67*, 37–44. [CrossRef]
144. Soto, D.; Zúñiga, L. Zooplankton assemblages of Chilean temperate lakes: A comparison with North American counterparts. *Rev. Chil. Hist. Nat.* **1991**, *64*, 569–581.

145. De los Ríos, P.; Crespo, J. Salinity effects on the abundance of *Boeckella popoensis* (Copepoda, Calanoidea) in saline ponds of the Atacama Desert, northern Chile. *Crustaceana* **2004**, *77*, 417–423.
146. Soto, D.; De los Rios, P. Trophic status and conductivity as regulators of daphnids dominance and zooplankton assemblages in lakes and ponds of Torres del Paine National Park. *Biol. Bratisl.* **2006**, *61*, 541–546. [CrossRef]
147. De los Ríos, P.; Soto, D. Crustacean (Copepoda and Cladocera) zooplankton richness in Chilean Patagonian lakes. *Crustaceana* **2007**, *80*, 285–296. [CrossRef]
148. Mühlhauser, H.; Hrepic, N.; Mladinic, P.; Montecino, V.; Cabrera, S. Water quality and limnological features of a high-elevation Andean Lake, Chungará in northern Chile. *Rev. Chil. Hist. Nat.* **1995**, *68*, 341–349.
149. Derry, A.; Prepas, E.; Hebert, P. A comparison of zooplankton communities in saline lake water with variable anion composition. *Hydrobiologia* **2003**, *505*, 199–215. [CrossRef]
150. Battauz, Y.; José de Paggi, S.; Romano, M.; Barbaeris, I. Zooplankton characterization of the Pampean saline shallow lakes, habitat of the Andean Flamingoes. *J. Limnol.* **2013**, *72*, 531–542. [CrossRef]

Article

Drought Monitoring over West Africa Based on an Ecohydrological Simulation (2003–2018)

Hiroyuki Tsutsui [1,*], Yohei Sawada [2], Katsuhiro Onuma [1], Hiroyuki Ito [1] and Toshio Koike [1]

[1]   International Centre for Water Hazard and Risk Management (ICHARM), Public Works Research Institute (PWRI), 1-6, Minamihara, Tsukuba 305-8516, Japan; oonuma-k573ck@pwri.go.jp (K.O.); h-itou@pwri.go.jp (H.I.); koike@icharm.org (T.K.)

[2]   Institute of Engineering Innovation, The University of Tokyo, Tokyo 113-8656, Japan; yohei.sawada@sogo.t.u-tokyo.ac.jp

*   Correspondence: t-tsutsui55@pwri.go.jp; Tel.: +81-29-879-6779

Abstract: In Africa, droughts are causing significant damage to human health and the economy. In West Africa, a severe decline in food production due to agricultural droughts has been reported in recent years. In this study, we simulated ecohydrological variables using the Coupled Land and Vegetation Data Assimilation System, which can effectively evaluate the hydrological water cycle and provide a dynamic evaluation of terrestrial biomass. Using ecohydrological variables (e.g., soil moisture content, leaf area index and vegetation water content) as a drought indicator, we analyzed agricultural droughts in the Sahel-inland region of West Africa during 2003–2018. Results revealed reasonable agreement between the simulated values and the pearl millet yield, and produced a successful quantification of severe droughts in the Sahel-inland region.

Keywords: drought; West Africa; ecohydrology; data assimilation; microwave remote sensing; vegetation water content; soil moisture; locust plague

Citation: Tsutsui, H.; Sawada, Y.; Onuma, K.; Ito, H.; Koike, T. Drought Monitoring over West Africa Based on an Ecohydrological Simulation (2003–2018). *Hydrology* **2021**, *8*, 155. https://doi.org/10.3390/hydrology8040155

Academic Editor: Luca Brocca

Received: 6 September 2021
Accepted: 11 October 2021
Published: 14 October 2021

Publisher's Note: MDPI stays neutral with regard to jurisdictional claims in published maps and institutional affiliations.

1. Introduction

In Africa, floods and droughts have become a serious issue. The number of flood events is increasing considerably year-to-year, and the occurrence of drought is causing both substantial economic damage and harm to human health (source: Munich Re Nat-CatSERVICE: https://www.iii.org/graph-archive/96134 accesed on 30 September 2021 and [1]). Furthermore, it is predicted that climate change will cause severe floods and droughts to occur more frequently in the future in land areas within the monsoon domains of North Africa [2]. Additionally, the proportion of the population with access to at least basic drinking water services in 2015 was much lower in sub-Saharan Africa than in the other regions of the world, and 58% of the regional population had no alternative to collecting untreated and often contaminated drinking water directly from surface water sources [3]. Overall, the level of development of basic sanitation infrastructure is <50% in almost all countries within this region [3]. Moreover, this region was the only region during 1990–2013, that registered an increase in the absolute number of people living in extreme poverty [3]. In these circumstances, the occurrence of droughts or floods can be highly detrimental to food security. Africa has an issue with water-related disasters (particularly flood and drought disasters) and their consequences regarding socioeconomic development. Resolving this dire situation will require development of a system for data integration, information fusion, synthesis, information sharing and communication promotion. With such a system, it would be possible to ensure the maximum use of data and information from observation, monitoring, prediction, and socioeconomic surveys and statistics, which can assist African countries in overcoming such problems. In West Africa, the impact of agricultural droughts is becoming increasingly evident because of the Charny effect that links surface albedo and precipitation [4], i.e., an increase in albedo due to reduced vegetation coverage causes precipitation to decrease, which results in a further increase in albedo because more of the land surface will be exposed owing to diminished growth of vegetation.

In West Africa, many researchers have been studying droughts, and the findings of some have identified the importance of (1) the contrast between the pre-monsoon and peak monsoon seasons, (2) two preferred modes of interannual variability (a latitudinal displacement of the tropical rain belt and changes in its intensity, and (3) the tropical easterly jet [5]. Moreover, other research identified notable trends of decrease in rainfall in the Sahel region (10–20° N, 18°W–20° E) from the late 1950s to the late 1980s. Although, Sahel rainfall recovered somewhat through to 2003, drought conditions within the region did not end [6]. These earlier approaches to drought assessment focused on the investigation of monsoon and rainfall trends using conventional drought indexes. Drought prediction over West Africa using the Standardized Precipitation Evapotranspiration Index (SPEI) and the Standardized Precipitation Index (SPI) has been implemented under the RCP4.5 and RCP8.5 scenarios [7]. In a study of Niger River Basin in West Africa, the performances of three drought indexes, i.e., the Standardized Rainfall Anomaly Index, the Bhalme and Mooley Drought Index, and SPI, were evaluated and compared [8]. Additionally, the question of how rising global temperatures might affect the spatial pattern of rainfall and the resultant droughts in West Africa was also investigated. Furthermore, precipitation and potential evapotranspiration variables have been simulated using the Rossby Centre RCA4 regional atmospheric model driven by 10 global climate levels under the RCP8.5 scenario (CanESM2, CNRM-CM5, CSIRO-Mk3, EC-EARTH-r12, GFDL-ESM2M, HadGEM2-ES, IPSL-CM5A-MR, MIROC5, MPI-ESM-LR, and NorESM1-M) [9]. This approach to drought assessment is based on the use of atmospheric and land surface models.

The mainstream approach to the assessment of drought in West Africa has concentrated on monsoons and rainfall trends [5,6] using conventional drought indexes [7,8], e.g., the SPI, SPEI and the self-calibrating Palmer Drought Severity Index. Although drought assessment using conventional drought indexes can be effective, the emphasis of such an approach is placed on precipitation, with limited consideration of land surface hydrology and energy circulation. In studies using satellite remote sensing, the vegetation condition (e.g., the normalized difference vegetation index (NDVI), leaf area index (LAI), and vegetation optical depth (VOD)) is monitored using visible and near-infrared sensors such as the Moderate Resolution Imaging Spectroradiometer. Furhermore, the near-surface soil moisture content can be monitored using passive and active microwave sensors such as the Advanced Microwave Scanning Radiometer for the Earth Observing System (AMSR-E), Advanced Microwave Scanning Radiometer 2 (AMSR2), the Soil Moisture Active Passive satellite and the Soil Moisture and Ocean Salinity satellite. Thus, although satellite remote sensing cannot be used to assess the root-zone soil moisture content, which is important for vegetation growth dynamics, it can be used to evaluate the near-surface soil moisture content. Therefore, water absorption from the root-zone layer and vegetation growth dynamics are not considered in the NDVI, LAI and VOD. The Global Land Data Assimilation System (GLDAS) [10] is the land surface data assimilation system integrated between a land surface model and a data assimilation scheme, in which the skin temperature is assimilated. In the GLDAS, not only the near-surface soil moisture content but also the root-zone soil moisture content is calculated by assimilating the skin temperature which is a variable of land surface hydrology and energy circulation used in calculating the flux by the land surface model. For vegetation, NDVI and LAI are used on the basis of visible and near infrared remote sensing. The Coupled Land and Vegetation Data Assimilation System (CLVDAS) [11–13] integrates passive microwave remote sensing techniques, a land surface model, a dynamic vegetation model (DVM) and a data assimilation scheme. CLVDAS assimilates microwave brightness temperatures (6.925 GHz; vertical polarization, 6.925 GHz; horizontal polarization, 10.7 GHz; vertical polarization and 10.7 GHz; horizontal polarization) that represent the product between the skin temperature and microwave emissivity. Thus, CLVDAS can estimate not only the optimized near-surface soil moisture content but also the optimized root-zone soil moisture content through data assimilation, because the microwave brightness temperature is sensitive to moisture. Additionally, the integrated DVM can be used to estimate the optimized LAI and vegetation water content, which can provide evaluation of vegetation growth dynamics, through data assimilation of microwave brightness temperatures. Thus, the gap between conventional study and CLVDAS-based studies is as follows: (1) evaluation of optimized root-zone soil moisture content through data assimilation, and (2) evaluation of optimized LAI and vegetation water content based on vegetation growth dynamics.

To fill this gap, land surface hydrology and energy circulation were evaluated using CLVDAS [11–13] in this study. Because this approach provides the possibility of drought monitoring and application to agricultural support [14], we evaluated the relationship between the pearl millet yield as a major crop and the simulated vegetation water content as a drought index, and analyzed the applicability of the approach to the assessment of agricultural droughts in the Sahel-inland region of West Africa during 2003–2018.

2. Data

CLVDAS needs global meteorological forcing data, such as precipitation (mm/s), air temperature (K), air pressure (mbar), shortwave radiation (W/m^2), longwave radiation (W/m^2), wind speed (m/s) and specific humidity (kg/kg), for EcoHydro-SiB and the assimilation of global satellite-observed microwave brightness temperature data for data. The suitability for CLVDAS of GLDAS ver. 2.1 global meteorological forcing data has been recognized [11–14] and therefore the GLDAS global meteorological forcing data, which can be downloaded from https://urs.earthdata.nasa.gov accesed on 30 September 2021, were used in this study. The GLDAS meteorological data have 3-h temporal resolution and $0.25° \times 0.25°$ gridded spatial resolution, i.e., the same as the output of CLVDAS. Satellite-observed microwave brightness temperatures (vertical and horizontal polarizations at 6.925, 10.65, and 18.7 GHz) from the AMSR-E and AMSR2, which can be downloaded from https://gportal.jaxa.jp/gpr/?lang=en accesed on 30 September 2021, were also used. These data have daily temporal resolution (descending orbit only), but $0.25° \times 0.25°$ gridded spatial resolution, i.e., the same as the output of CLVDAS. The period from October 2011 to December 2012 represents the period of transition from AMSR-E to AMSR2, and data assimilation was not conducted because microwave brightness temperatures were not observed by the satellites. Therefore, the ecohydrological variable was not provided by CLVDAS during this period (Table 1). Crop yield data, obtained from the Food and Agriculture Organization of the United Nations, were downloaded from http://faostat3.fao.org/download/Q/QC/E accesed on 30 September 2021.

Table 1. List of input and output datasets.

	Items	Unit	Source	Spatial Resolution	Temporal Resolution	Regions	Periods in This Study
	Precipitation	mm/s	GLDAS 2.1	0.25°	3 h	Global	2003.1–2018.12
	Air temperature	K	GLDAS 2.1	0.25°	3 h	Global	2003.1–2018.12
	Air pressure	mbar	GLDAS 2.1	0.25°	3 h	Global	2003.1–2018.12
	Shortwave radiation	W/m^2	GLDAS 2.1	0.25°	3 h	Global	2003.1–2018.12
	Longwave radiation	W/m^2	GLDAS 2.1	0.25°	3 h	Global	2003.1–2018.12
	Wind speed	m/s	GLDAS 2.1	0.25°	3 h	Global	2003.1–2018.12
	Specific humidity	kg/kg	GLDAS 2.1	0.25°	3 h	Global	2003.1–2018.12
Input dataset	6.925 GHz microwave brightness temperature (V polarization)	K	AMSR-E, AMSR2	0.25°	1 day	Global	2003.1–2011.9 2013.1–2018.12
	6.925 GHz microwave brightness temperature (H polarization)	K	AMSR-E, AMSR2	0.25°	1 day	Global	2003.1–2011.9 2013.1–2018.12
	10.7 GHz microwave brightness temperature (V polarization)	K	AMSR-E, AMSR2	0.25°	1 day	Global	2003.1–2011.9 2013.1–2018.12
	10.7 GHz microwave brightness temperature (H polarization)	K	AMSR-E, AMSR2	0.25°	1 day	Global	2003.1–2011.9 2013.1–2018.12

Table 1. Cont.

	Items	Unit	Source	Spatial Resolution	Temporal Resolution	Regions	Periods in This Study
Output dataset	Near-surface soil moisture contnet	m^3/m^3	CLVDAS	0.25°	1 day	West Africa	2003.1–2018.12
	Root-zone soil moisture contnet	m^3/m^3	CLVDAS	0.25°	1 day	West Africa	2003.1–2018.12
	Vegetation water content	m^3/m^3	CLVDAS	0.25°	1 day	West Africa	2003.1–2018.12

3. Methods

To assess agricultural droughts in West Africa, a methodology for estimating ecohydrological variables such as soil moisture content and vegetation water content is described.

3.1. Study Area and Period

The selected simulation domain of West Africa comprised the region 0°30′–25°7′ N, 18°7′ W–16°7′ E. The Sahel-inland region (10° N–16° N, 12° W–16° E), which is an active agricultural area, was selected as the specific study area for the assessment of agricultural droughts during 2003–2018 (Figure 1).

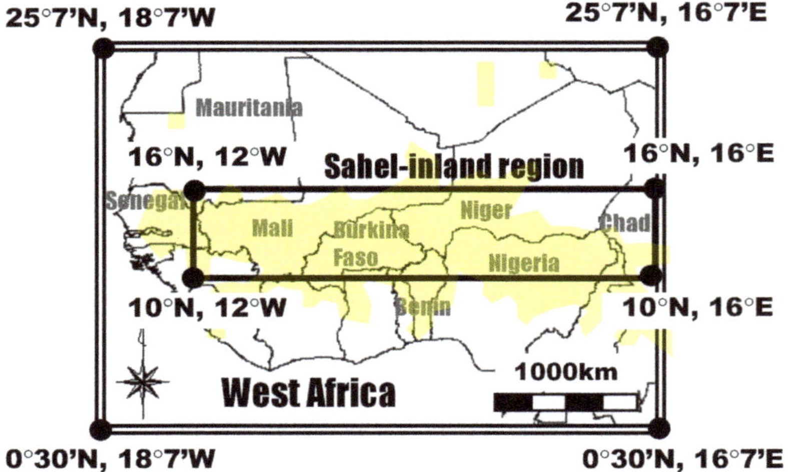

Figure 1. Simulation domain (0°30′–25°7′ N, 18°7′ W–16°7′ E) and study area (10°–16° N, 12° W–16° E) in West Africa. Agricultural drought assessment was conducted for the Sahel-inland region; yellow shading indicates the general area of pearl millet cropland [15].

3.2. System Overview

As listed in Table 1 and illustrated in Figure 2, this research used CLVDAS [11,12] to calculate the ecohydrological variables. Meteorological forcing data (Figure 2a) are input to EcoHydro-SiB (Figure 2b), which is a land surface model that can calculate various ecohydrological variables (Figure 2c). EcoHydro-SiB (Figure 2b) is coupled with Hydro-SiB and the dynamic vegetation model (DVM). The Simple Biosphere Model 2 (SiB2) [16] was improved, Hydro-SiB was developed based on a one-dimensional Richards's equation, and the vertical interlayer flows within the unsaturated zone [17]. Soil water dynamics are described by van Genuchten's water retention curve [18], and the LAI and vegetation are calculated on the basis of carbon-pool dynamics. For a detailed explanation and the formulations, the reader is referred to Section 2.1 [11]. The calculated ecohydrological variables (Figure 2c) are used to drive a microwave radiative transfer model (RTM; Figure 2d) to calculate microwave brightness temperatures (Figure 2e). The RTM (Figure 2d) combines the advanced integral equation model (AIEM) with a shadowing effect [19] to evaluate land surface scattering and the omega-tau model [20], which evaluates the microwave radiative transfer process in the ground surface. The cost is calculated on the basis of the difference between the calculated microwave brightness temperature (Figure 2e) and the satellite observed microwave brightness temperature (Figure 2f) of the land surface (Equation (1)). The calculated cost (Figure 2g) is minimized through data assimilation.

$$Cost = \sum_{F=6,10GHz} \sum_{P=H,V} \left(TBe_F^P - TBo_F^P \right)^2, \tag{1}$$

where TBe_F^P is the calculated microwave brightness temperature, TBo_F^P is the satellite observed microwave brightness temperature, and F and P are frequency (GHz) and polarization (H-horizontal, V- vertical) respectively.

Using the above methodology, the optimized ecohydrological variables (Figure 2j), e.g., near-surface soil moisture content, root-zone soil moisture content, evapotranspiration, LAI, and vegetation water content, can be estimated. By assimilating the satellite microwave brightness temperatures of the land surface, it is possible to estimate all ecohydrological variables spatiotemporally on the global scale. This system has two different modules: parameter optimization and data assimilation. In the parameter optimization module (Figure 2h), the shuffled complex evolution (SCE as a data assimilation scheme) [21] (Figure 2i), determines the most important optimized parameter. In the data assimilation module, the genetic particle filter (GPF as a data assimilation scheme) [22], estimates optimized ecohydrological variables (Figure 2i) sequentially.

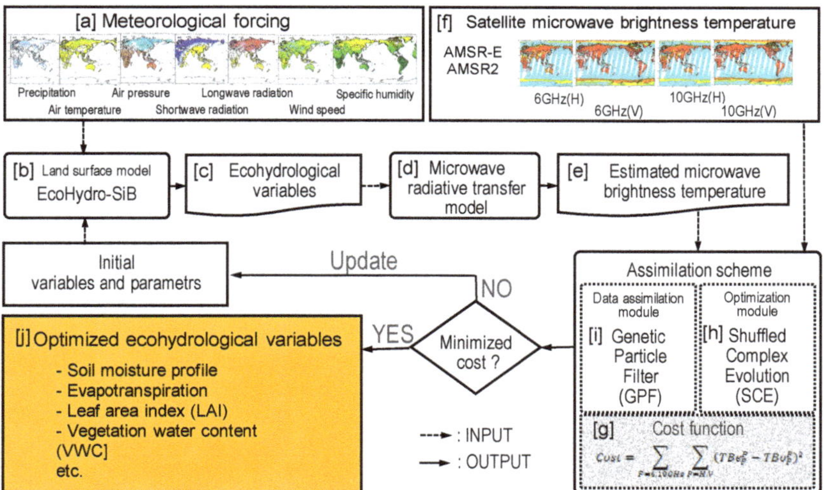

Figure 2. CLVDAS framework used in this study. Optimized ecohydrological variables are outputted from this system, and the near-surface soil moisture content, root-zone soil moisture content, and the vegetation water content, as optimized ecohydrological variables, are analyzed. The meteorological forcing dataset comprises precipitation (mm/s), air temperature (K), air pressure (mbar), shortwave radiation (W/m^2), longwave radiation (W/m^2), wind speed (m/s), and specific humidity (kg/kg). Satellite-observed brightness temperature dataset comprises brightness temperatures for 6.925 GHz horizontal polarization (6 GHz(H)), 6.925 GHz vertical polarization (6 GHz(V)), 10.7 GHz horizontal polarization (10 GHz(H)), and 10.7 GHz vertical polarization (10 GHz(V)). Meteorological forcing data (**a**) are input to EcoHydro-SiB (**b**). EcoHydro-SiB is a land surface model that calculates various ecohydrological variables (**c**). The calculated ecohydrological variables (**c**) are used to drive a microwave

radiative transfer model (**d**) to calculate microwave brightness temperatures (**e**). The cost is calculated on the basis of the difference between the calculated microwave brightness temperatures (**e**) and the satellite-observed microwave brightness temperatures (**f**) of the land surface. The calculated cost (**g**) is minimized through data assimilation scheme of the Shuffled Complex Evolution (SCE) (**h**) and the Genetic Particle Filter (GPF) (**i**). Ultimately, the optimized eco-hydrological variables (**j**) are estimated.

3.3. Drought Index

In this study, the near-surface soil moisture content (0–3 cm depth, m^3/m^3) [11,12], root-zone soil moisture content (3–20 cm depth, m^3/m^3), and vegetation water content (m^3/m^3) were estimated using CLVDAS to investigate the ecohydrological water cycle and agricultural drought. This study considered that by absorbing sufficient water from the roots, crops can store ample water in the plant body, grow well, and bear much fruit. Therefore, this study focused on vegetation water content (m^3/m^3) as an indicator of agricultural droughts. In West Africa, the cultivation of rain-fed crops for domestic consumption is widespread and pearl millet represents the principal staple crop. Therefore, pearl millet was selected as another indicator of agricultural drought in this study. In West Africa, pearl millet is sown during June–July, grows during August–September, and is harvested after October. Therefore, September, representing the period of maximum growth to the fruiting period, is an important time in which to assess agricultural drought. Thus, the vegetation water content in September (temporal average) and the pearl millet crop yield were also selected as drought indicators in this study. Because it is not possible to compare various ecohydrological variables, such as soil moisture content, vegetation water content, and crop yield quantitatively, the normalized index (NI_i) based on the z-score theory was calculated for each day from 2003 to 2018 using Equation (2):

$$NI_i = \frac{x_i - \mu}{\sigma}, \tag{2}$$

where x_i is a variable (i.e., near-surface soil moisture content, root-zone soil moisture content, vegetation water content, pearl millet yield, and number of days with a locust outbreak) on arbitrary date (i) in a year, and μ and σ are the average and standard deviation for x_i on arbitrary date (i) in all years (2003–2018). Values of x_i for near-surface soil moisture content, root-zone soil moisture content, and vegetation water content were calculated for each grid by CLVDAS, as shown in Figure 3. Therefore, μ and σ were also calculated for each grid. Using these values of x_i, μ, and σ, the normalized index (NI_i) based on the z-score theory was calculated for each grid, as shown in Figure 4. Subsequently, the normalized index (NI_i) values were averaged spatially for the Sahel-inland region. Finally, the normalized index (NI_i) values were averaged temporally for September, as shown in Figure 5. Values of x_i for the pearl millet crop yield represent the annual total yield in the Sahel-inland region (i.e., Chad, Niger, Nigeria, Benin, Burkina Faso, and Mali) in all years (2003–2018). Therefore, μ and σ were calculated using the x_i value for each year. Finally, the annual normalized index (NI_i) values for the pearl millet crop yield were calculated, as shown in Figure 6. The number of days with a locust outbreak were calculated as follows. The number of days (x_i) in each year (2003–2018) were counted when more than 10 locust plagues occurred in the Sahel-inland region, according to the Locust watch of the Food and Agriculture Organization (FAO) of the United Nations. Then, the average (μ) and standard deviation (σ) were calculated using these numbers (x_i) for each year. Using these values of x_i, μ, and σ, the normalized index (NI_i) based on the z-score theory was calculated, as shown in Figure 7.

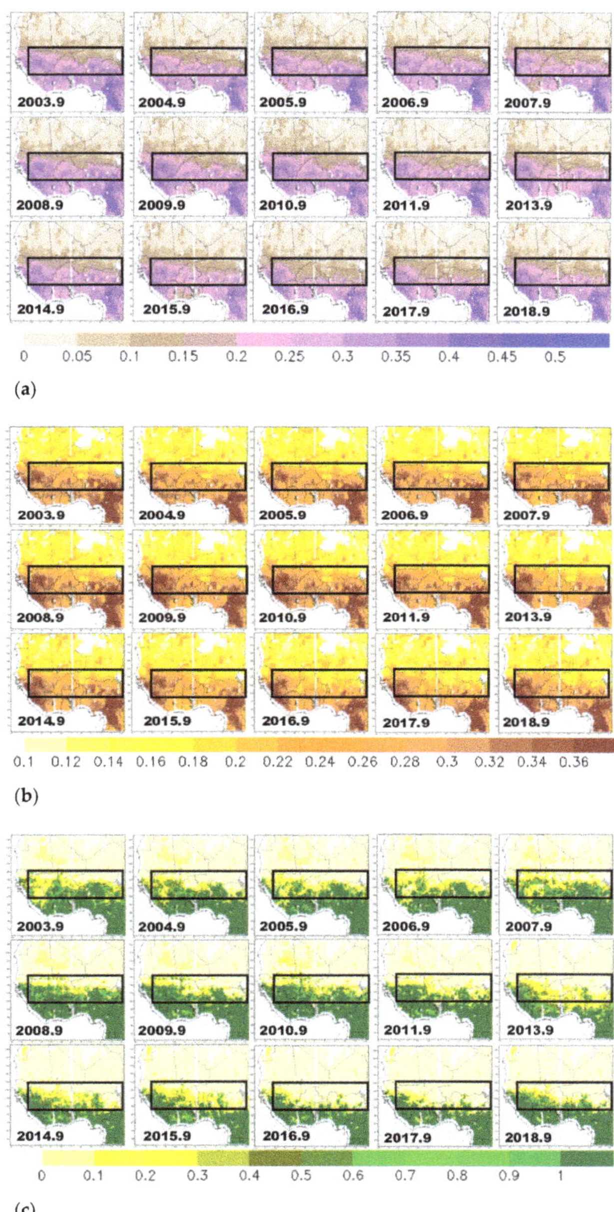

Figure 3. Spatial distribution of (**a**) near-surface soil moisture content (m^3/m^3), (**b**) root-zone soil moisture content (m^3/m^3), and (**c**) vegetation water content (m^3/m^3) from CLVDAS (September monthly averages in the period 2003–2018; spatial resolution is 0.25° × 0.25°). Black rectangle outlines the Sahel-inland region. In the Sahel-inland region, vegetation water content has shown a trend of decrease from the north since 2011 (**c**). This trend has also shown in the root-zone soil moisture content (**b**) although not clearly in the near-surface soil moisture content (**a**).

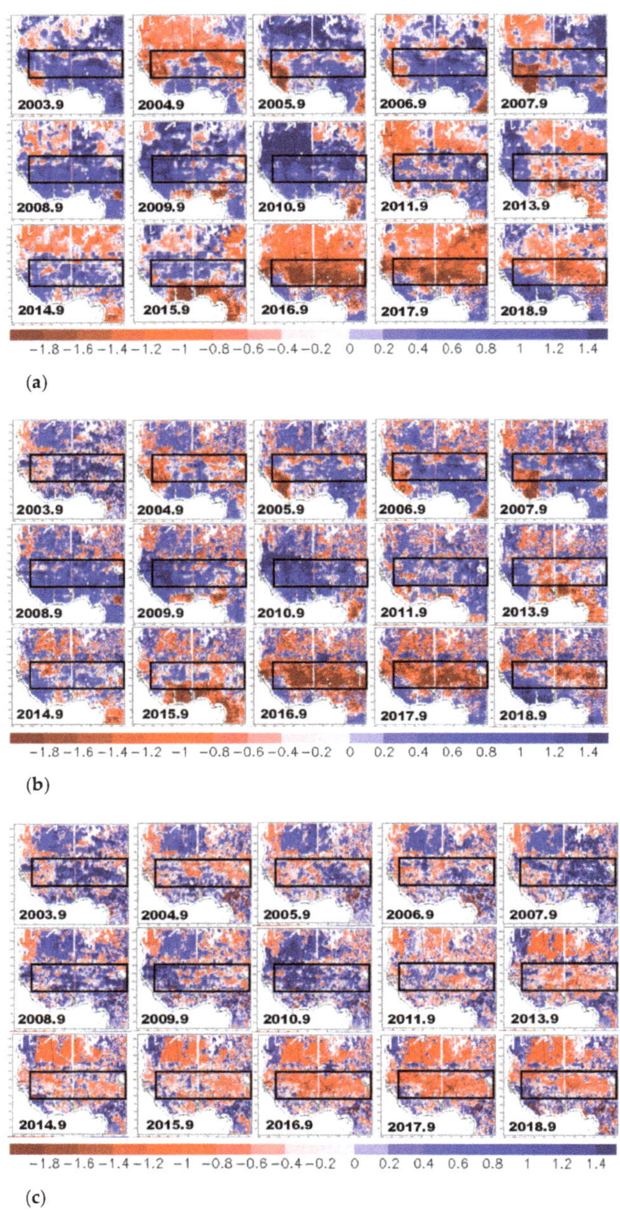

Figure 4. Spatial distribution of NI_i for (**a**) near-surface soil moisture content (m³/m³), (**b**) root-zone soil moisture content (m³/m³), and (**c**) vegetation water content (m³/m³) from the CLVDAS (September monthly averages in the period 2003–2018; spatial resolution is 0.25° × 0.25°). Black rectangle outlines the Sahel-inland region. Although variation in vegetation water content itself is slightly unclear (Figure 3c), its normalized index (NI_i) based on the z-score theory clearly shows the decrease since 2011(**c**). Furthermore, this trend is also shown in the near-surface and root-zone soil moisture content (**a**) and (**b**), respectively.

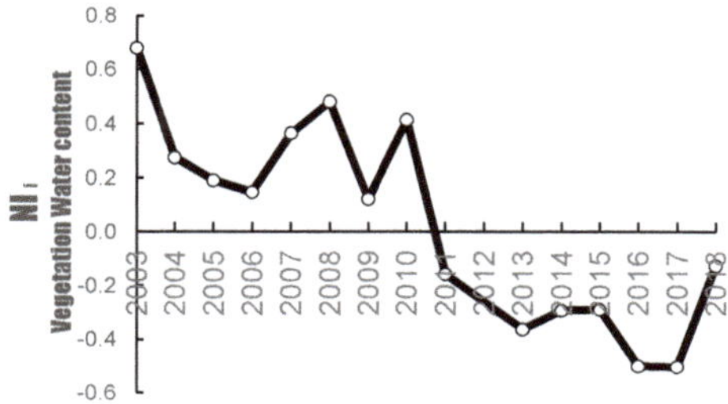

Figure 5. Annual variation in NI_i for vegetation water content in the Sahel-inland region (September monthly averages in the period 2003–2018; spatial average of entire Sahel-inland region: resolution is $0.25° \times 0.25°$). A substantial decrease in the normalized index (NI_i) of the vegetation water content since 2010 is shown.

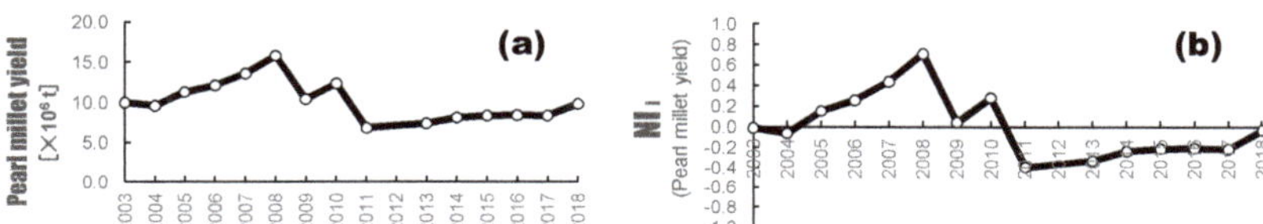

Figure 6. Annual variation in the pearl millet yield in the Sahel-inland region consisting of Chad, Niger, Nigeria, Benin, Burkina Faso, and Mali: (**a**) pearl millet yield, and (**b**) its normalized index (NI_i), decrease in pearl millet yield since 2011 is clearly shown.

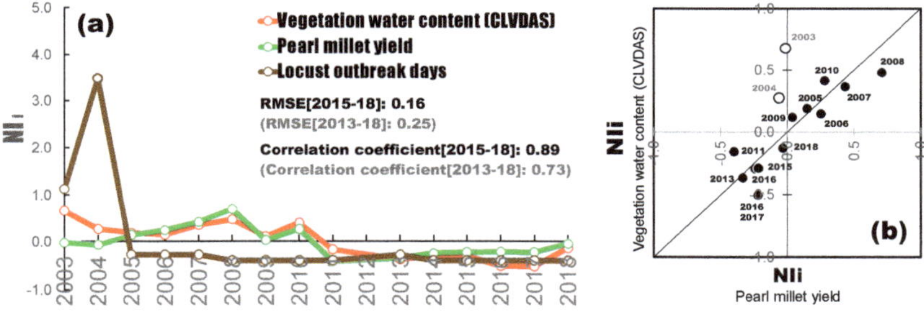

Figure 7. (**a**) Comparison of the normalized index (NI_i) for the vegetation water content (CLVDAS) (blue line), pearl millet yield (red line), and number of locust outbreak days (brown line), and (**b**) scatterplot between the NI_i for vegetation water content and the NI_i for pearl millet yield of the Sahel-inland region. The number of locust outbreak days was calculated as follows. The number of days (x_i) in each year (2003–2018) were counted when more than 10 locust plagues occurred in the Sahel-inland region according to the FAO Locust watch. Then, the average (μ) and standard deviation (σ) were calculated using these numbers (x_i) for each year. Using these values of x_i, μ, and σ, the normalized index (NI_i) based on the z-score theory was calculated. In this study, we assessed agricultural drought for the period 2005–2018, excluding 2003–2004. For this period, the RMSE between the NI_i for pearl millet yield and the NI_i for vegetation water content was 0.16 and the correlation coefficient was 0.89, indicating strong agreement.

4. Results and Discussion

In passive microwave remote sensing, the low-frequency band with the longest wavelength is used by AMSR-E and AMSR2 because the microwaves emitted from the soil must be scanned through the atmosphere and vegetation. However, a passive microwave sensor can generally detect only the near-surface soil moisture content because even microwaves in the low-frequency band are absorbed by soil moisture [23]. This is the reason why the AMSR-E and AMSR2 soil moisture product targets only the near-surface soil moisture content [23]. To overcome this shortcoming, as described in Section 1, the following processing is implemented in CLVDAS. Eco-HydroSiB evaluates the ecohydrological water cycle, which expresses the penetration of precipitation, water storage in the root-zone, and water absorption by roots and vegetation growth, and simulates the ecohydrological variables (e.g., soil moisture profile, evapotranspiration, and biomass). Furthermore, the RTM calculates the microwave brightness temperatures emitted from the land surface using the simulated ecohydrological variables. Additionally, changes in the ecohydrological variables and assimilation of the microwave brightness temperatures are repeated until the difference between the simulated and satellite-observed microwave brightness temperatures of the ground surface is minimized. Hence, an accurate estimation of the ecohydrological water cycle is derived using this process. This represents the major advantage of this system. Hitherto, the CLVDAS outputs of estimated soil moisture content and LAI were validated by comparison with the following observations: ground-based observed soil moisture content and LAI observed at the Yanco Flux Tower site located in New South Wales (Australia) [12], ground-based observed soil moisture content from the African Monsoon Multidisciplinary Analyses, Vaira Ranch (USA), Bayantsagaan (Mongolia) [11,13], and the Moderate Resolution Imaging Spectroradiometer LAI in West Africa and Northeast Brazil [14]. The following estimation accuracy was achieved: (1) root mean square error (RMSE) of 0.05 m^3/m^3 or less and bias of 0.045 m^3/m^3 or less in terms of soil moisture content, and (2) RMSE of 0.12 m^2/m^2 or less and bias of 0.14 m^2/m^2 or less in terms of LAI.

For the period from 1 January 2003to 31 December 2018, the near-surface soil moisture content (m^3/m^3), root-zone soil moisture content (m^3/m^3), and vegetation water content (m^3/m^3) were simulated using CLVDAS and used to create a gridded dataset (temporal resolution: daily, spatial resolution: $0.25° \times 0.25°$). Figure 3 shows the spatial distribution of the averaged ecohydrological variable in September. In the Sahel-inland region, vegetation water content has shown a trend of decrease from the north since 2011 (Figure 3c). This trend has also shown in the root-zone soil moisture content (Figure 3b) but not so clearly in the near-surface soil moisture content (Figure 3a). Furthermore, NI_i for each ecohydrological variable was calculated for each grid using Equation (2). Figure 4 shows the spatial distribution of each normalized index of the averaged ecohydrological variables in September. Although variation in the vegetation water content itself is slightly unclear (Figure 3c), its normalized index (NI_i) based on the z-score theory clearly shows a decrease since 2011 (Figure 4c). Furthermore, this trend has also shown in the near-surface and root-zone soil moisture content (Figure 3a,b). Figure 5 shows the annual variation in the normalized index (NI_i) based on the z-score theory for vegetation of water content in the Sahel-inland region, calculated using Equation (2) (September average in the agricultural drought assessment period from 2003 to 2018, regional spatial average). We calculated the annual total yield of pearl millet, in the Sahel-inland region consisting of Chad, Niger, Nigeria, Benin, Burkina Faso, and Mali from 2003 to 2018 (Figure 6a), which revealed that the annual yield since 2011 has been approximately half that in 2008. Furthermore, we calculated the normalized index (NI_i) based on the z-score theory of the total yields of pearl millet using Equation (2) (Figure 6b). Figure 7 shows the annual variation in NI_i for both pearl millet yield and vegetation water content in the Sahel-inland region (temporal average in September and regional spatial average). The NI_i values for the pearl millet yield are negative during 2003–2004, whereas the concurrent NI_i values for vegetation water content are positive, and the difference between the two sets of NI_i values is large. We investigated external factors other than droughts by considering previous studies [24,25] and the FAO Locust watch (http://www.fao.org/ag/locusts/en/archives/briefs/index.html accessed on 30 September 2021), which revealed that serious locust outbreaks occurred during 2003–2004 in West Africa. The number of days (x_i) in each year (2003–2018) was counted when more than 10 locust plagues occurred in the Sahel-inland region using the FAO Locust watch. The average (μ) and standard deviation (σ) were calculated using these the numbers of days (x_i) in each year. Using x_i, μ, and σ, the normalized index (NI_i) based on the z-score theory is calculated by using Equation (2) (brown line). The NI_i values for locust outbreaks (brown line in Figure 7; calculated as described in Section 3.3) in 2003 and 2004 are 1.31 and 3.37, respectively; although, all NI_i values after 2005 are negative. We recognize that crop yields of the Sahel-inland region were likely to be adversely affected by the external impact of locust plagues in 2003–2004. Therefore, we assessed agricultural drought in the period from 2005 to 2018. For this period, the RMSE between the NI_i for pearl millet yield (green line in Figure 7a) and the NI_i for vegetation water content (red line in Figure 7a) was 0.16 and the correlation coefficient was 0.89, indicating a strong agreement. As an aside, the RSME and correlation coefficient values when including the period 2003–2004 were 0.25 and 0.73, respectively).

The variation in NI_i for precipitation and the simulated ecohydrological variables (near-surface soil moisture content, root-zone soil moisture content and vegetation water content) were investigated for the Sahel-inland region during the agricultural drought assessment period (2005–2018) (Figures 8 and 9). In the first half of the agricultural drought assessment period (Figure 8), the NI_i of vegetation water content was mostly positive, except in 2005, 2008 and 2009, which indicates that the effect of drought on vegetation water content is small. We also found that each peak had a time lag when focusing on the negative peak for precipitation, soil moisture content, and vegetation water content in 2005, 2008, and 2009 (yellow marks and lines in Figure 8). This is attributable to the following process: (i) the land surface soon dries because of the shortage of precipitation, (ii) the root-zone soil moisture decreases after a further amount of time because water is not supplied from the land surface, and (iii) vegetation water content declines after an even further amount of time because of the lack of root-zone soil moisture available for absorption by roots. In the second half of the agricultural drought assessment period (Figure 9), the vegetation water content gradually became negative after 2014. As in the first half of the agricultural drought assessment period (Figure 8), the shortage of precipitation gradually propagated to the lack of vegetation water content (yellow marks and lines in Figure 9). Although the NI_i values of precipitation became positive in April 2016, negative values remained in the growth and early harvest seasons (May–October), leading to negative NI_i values of soil moisture and vegetation water content simultaneously (green line in Figure 9). The NI_i values of precipitation became positive in the second half of December 2016 owing to the occurrence of heavy rainfall; however, they reverted to negative values in the beginning of January 2017. The surface soil moisture content showed the same behavior. In contrast, the root-zone soil moisture content was positive in the period from January–April because the water associated with the heavy rainfall was stored in the rootzone. The vegetation grew by absorbing the stored root-zone soil moisture, as indicated by the positive peak of vegetation water content at the beginning of April (blue line in Figure 9). This indicates that storage of root-zone soil moisture is important for vegetation growth, and that CLVDAS can evaluate such a mechanism. In 2017, soil moisture was not stored in the near-surface soil because precipitation amounts were low. Following subsequent rainfall events, the NI_i values of near-surface soil moisture recovered (became positive) in March 2018; however, the NI_i values of root-zone soil moisture and vegetation water content did not recover (remained negative). Thus, both the root-zone soil moisture content and vegetation water content remained negative in the long term because they have a long retention period of the memory of past water shortages (red line in Figure 9). By investigating the daily variation in precipitation, soil moisture content, and vegetation water content using CLVDAS, we were able to evaluate the following land surface hydrological water cycle and vegetation growth dynamics mechanism. (i) Reduced precipitation causes aridity of the land surface, which affects the condition of the root-zone soil moisture. (ii) Plants cannot retain sufficient water within their structures because of the lack of soil moisture available for absorption in the root-zone layer. (iii) Near-surface soil moisture content can change rapidly in response to temporary rainfall events. In contrast, both the root-zone soil moisture and the vegetation water content tend to remain in long-term drought conditions because they have a long retention period of the memory of past water shortage. The major finding of this study was establishing that CLVDAS can be used to evaluate the hydrological water cycle (penetration of precipitation to the root-zone soil layer and its absorption by roots) and vegetation growth dynamics. Additionally, we confirmed that vegetation water content output by CLVDAS can be used to assess agricultural drought through comparison with major crop yields and investigation of external factors in the target region.

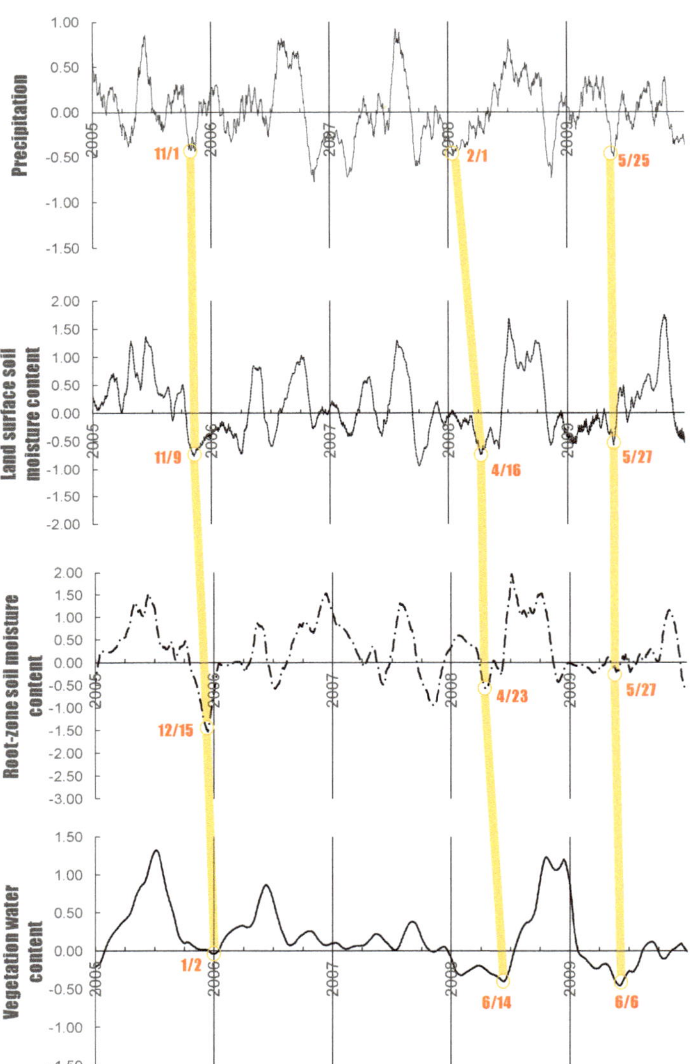

Figure 8. Normalized index (NI_i) for precipitation and ecohydrological variables in the first half of the agricultural drought evaluation period (2005–2009). Spatial averages for the Sahel-inland region. From the top, the graphs present precipitation, near-surface soil moisture content, root-zone soil moisture content, and vegetation water content. Dates in orange indicate the middle day of each negative peak period. We found that each peak had a time lag when focusing on the negative peak for precipitation, soil moisture content, and vegetation water content, as shown by the yellow marks and lines.

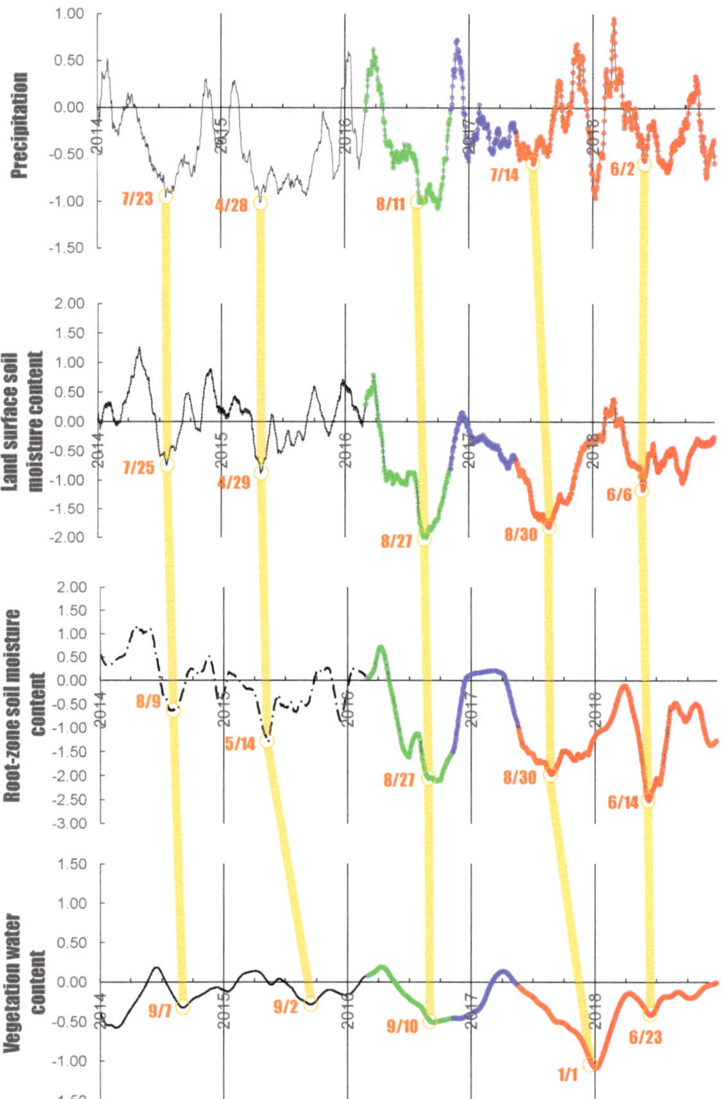

Figure 9. Normalized index (NI_i) for precipitation and ecohydrological variables in the second half of the agricultural drought evaluation period (2014–2018). Spatial averages for the Sahel-inland region. From the top, the graphs present precipitation, near-surface soil moisture content, root-zone soil moisture content, and vegetation water content. Dates in orange indicate the middle day of each negative peak period. We found that each peak had a time lag when focusing on the negative peak for precipitation, soil moisture content, and vegetation water content, as shown by the yellow marks and lines. As shown by the green lines, although the NI_i values of precipitation become positive in April 2016, negative values remain in the growth and early harvest seasons (May–October), which leads to negative NI_i values of soil moisture and vegetation water content simultaneously. As shown by the blue lines, although the NI_i values of precipitation became positive in the second half of December 2016 owing to the occurrence of heavy rainfall, they reverted to negative values in the beginning of January 2017. The surface soil moisture content showed the same behavior. In contrast,

the root-zone soil moisture content was positive in the period from January–April because the water associated with the heavy rainfall was stored in the root-zone. The vegetation grew by absorbing the stored root-zone soil moisture, as indicated by the positive peak of vegetation water content at the beginning of April. As shown by the red lines, both the root-zone soil moisture content and the vegetation water content remained negative in the long term because they have a long retention period of the memory of past water shortage.

5. Conclusions

To fill the gap of the conventional study as described in Section 1, this study used CLVDAS to simulate ecohydrological variables (particularly vegetation water content) for the use of drought indicators, and applied them to the analysis of drought in West Africa during 2013–2018. We found that the Sahel-inland region suffered locust plagues in 2003 and 2004. Because the impact of locust plagues on vegetation growth dynamics cannot be simulated by an ecohydrological model, we excluded the data for 2003 and 2004 from our analysis. The results of our agricultural drought assessment for the period 2005–2018 showed reasonable agreement (RMSE = 0.16 in the normalized index NI_i) between the pearl millet yield and the simulated vegetation water content in the Sahel-inland region. The strength of this agreement is attributable to the accurate simulation of the hydrological water cycle and vegetation growth dynamics by CLVDAS, which has great importance regarding agricultural drought assessment. Moreover, the possibility of identifying a relationship between precipitation a few months prior and crop yield was also suggested, although such a relationship also depends on the condition of water retention in the root-zone soil layer. These findings constitute the primary significance of conducting this study. One of the major limitations of the current application of CLVDAS is the spatial resolution of the CLVDAS output ($0.25° \times 0.25°$), which is slightly too wide owing to the assimilation of low-frequency microwave brightness temperatures with a wide observation footprint. Improvement of the spatial resolution of the CLVDAS gridded output is therefore an objective of our future work. Furthermore, the shortage of precipitation has considerable impact on the land surface condition in the rainy season of the following year, as was clarified by the CLVDAS simulation. The importance of deriving accurate initial conditions of soil moisture content and LAI in multi-seasonal drought prediction is therefore another area requiring improvement. In a previous study, a CLVDAS application involving a seasonal meteorological prediction from a general circulation model showed satisfactory performance in predicting the land surface conditions of a drought in the Horn of Africa [13]. Subsequently, a drought monitoring and seasonal prediction system based on CLVDAS was developed for Northeast Brazil [14]. This system can monitor and predict (three months) the soil moisture profile, evapotranspiration, and LAI. By coupling these previous studies with this study of West Africa, it is expected that not only vegetation water content but also crop yield can be predicted by simulating the conditions of the several previous months. Thus, developing CLVDAS for seasonal prediction and crop yield prediction will be addressed in future work.

Author Contributions: Conceptualization, resources, writing—review and editing, T.K., H.I., K.O., Y.S. and H.T.; methodology, T.K., Y.S. and H.T.; software, validation, Y.S.; formal analysis, investigation, data curation, writing—original draft preparation, visualization, H.T.; supervision, T.K., Y.S., H.I., K.O.; project administration, funding acquisition, T.K., H.I. and K.O. All authors have read and agreed to the published version of the manuscript.

Funding: This research was funded in part by the International Centre for Water Hazard and Risk Management (ICHARM)/Public Works Research Institute (PWRI),and the Data Integration and Analysis System project (DIAS-P) supported by the Ministry of Education, Culture, Sports, Science and Technology of Japan. JSPS KAKENHI grant JP17K18352 and 21H01430, and JAXA grant ER2GWF102.

Institutional Review Board Statement: Not applicable.

Informed Consent Statement: Not applicable.

Data Availability Statement: The Global Land Data Assimilation System version 2.1 (GLDAS 2.1) can be found at https://urs.earthdata.nasa.gov accessed on 30 September 2021. The AMSR-E and AMSR2 dataset can be downloaded from https://gportal.jaxa.jp/gpr/?lang=en accessed on 30 September 2021. The FAO Locust Watch dataset can found at http://www.fao.org/ag/locusts/en/archives/briefs/index.html accessed on 30 September. Crop yield data from FAOSTAT can be found at http://faostat3.fao.org/download/Q/QC/E accessed on 30 September 2021.

Acknowledgments: This research was conducted as part of a project led by the International Centre for Water Hazard and Risk Management (ICHARM)/Public Works Research Institute (PWRI) and also the Data Integration and Analysis System project (DIAS-P) supported by the Ministry of Education, Culture, Sports, Science and Technology of Japan. The Global Land Data Assimilation System version 2.1 (GLDAS 2.1) global meteorological forcing data and the passive microwave brightness temperature data (AMSR-E and AMSR2) were furnished by EARTHDATA of NASA and the JAXA G-portal, respectively. The information on agriculture and locust outbreaks was downloaded from the Food and Agriculture Organization of the United Nations. The authors thank the three anonymous reviewers for their helpful comments.

Conflicts of Interest: The authors declare no conflict of interest.

References

1. Quesada-Román, A. Deciphering Natural Hazard Histories Based on Tree-Ring Analyses in Contrasting Tropical Ecosystems of Costa Rica. Ph.D. Thesis, Université de Genève, Genève, Switzerland, 2020. [CrossRef]
2. IPCC. *2014: Climate Change 2014: Synthesis Report. Contribution of Working Groups I, II and III to the Fifth Assessment Report of the Intergovernmental Panel on Climate Change*; Core Writing Team, Pachauri, R.K., Meyer, L.A., Eds.; IPCC: Geneva, Switzerland, 2014; 151p.
3. UNESCO World Water Assessment Programme. *The United Nations World Water Development Report 2019*; UNESCO World Water Assessment Programme: Perugia, Italy, 2019; ISBN 978-92-3-100309-7.
4. Charney, J.; Stone, P.H.; Quirk, W.J. Drought in the Sahara: A biogeophysical feedback mechanism. *Science* **1975**, *187*, 434–435. [CrossRef]
5. Nicholson, S.E. The West African Sahel: A Review of Recent Studies on the Rainfall Regime and Its Interannual Variability. *ISRN Meteorol.* **2013**, *2013*, 1–32. [CrossRef]
6. Dai, A.; Lamb, P.J.; Trenberth, K.E.; Hulme, M.; Jones, P.D.; Xie, P. The Recent Sahel Drought Is Real. *Int. J. Climatol.* **2004**, *24*, 1323–1331. [CrossRef]
7. Ajayi, V.O.; Ilori, O.W. Projected Drought Events over West Africa Using RCA4 Regional Climate Model. *Earth Syst. Environ.* **2020**, *4*, 329–348. [CrossRef]
8. Okpara, J.N.; Tarhule, A. Evaluation of Drought Indices in the Niger Basin, West Africa. *J. Geogr. Earth Sci.* **2015**, *3*, 1–32. [CrossRef]
9. Quenum, G.M.L.D.; Klutse, N.A.B.; Dieng, D.; Laux, P.; Arnault, J.; Kodja, J.D.; Oguntunde, P.G. Identification of Potential Drought Areas in West Africa Under Climate Change and Variability. *Earth Syst. Environ.* **2019**, *3*, 429–444. [CrossRef]
10. Rodell, M.; Houser, P.R.; Jambor, U.; Gottschalck, J.; Mitchell, K.; Meng, C.-J.; Arsenault, K.; Cosgrove, B.; Radakovich, J.; Bosilovich, M.; et al. The Global Land Data Assimilation System. *Bull. Amer. Meteor. Soc.* **2004**, *85*, 381–394. [CrossRef]
11. Sawada, T.; Koike, T. Simultaneous estimation of both hydrological and ecological parameters in an ecohydrological model by assimilating microwave signal. *J. Geophys. Res. Atmos.* **2014**, *119*, 8839–8857. [CrossRef]
12. Sawada, Y.; Koike, T.; Walker, J.P. A land data assimilation system for simultaneous simulation of soil moisture and vegetation dynamics. *J. Geophys. Res. Atmos.* **2015**, *120*, 5910–5930. [CrossRef]
13. Sawada, Y.; Koike, T. Towards ecohydrological drought monitoring and prediction using a land data assimilation system: A case study on the Horn of Africa drought (2010–2011). *J. Geo-Phys. Res. Atmos.* **2016**, *121*, 8229–8242. [CrossRef]
14. Tsutsui, H.; Sawada, Y.; Ikoma, E.; Kitsuregawa, M.; Koike, T. Study on the estimation of crop production and required irrigation water based on long-term drought simulation by using the coupled land and vegetation data assimilation. *J. Jpn. Soc. Civ. Eng. Ser. B1* **2019**, *75*, I_283–I_288. [CrossRef]
15. Oumar, I.; Mariac, C.; Pham, J.; Vigouroux, Y. Phylogeny and origin of pearl millet (Pennisetum glaucum [L.] R. Br) as revealed by microsatellite loci. *Theor. Appl. Genet.* **2008**, *117*, 489–497. [CrossRef]
16. Sellers, P.J.; Randall, D.A.; Collatz, G.J.; Berry, J.A.; Field, C.B.; Dazlich, D.A.; Zhang, C.; Collelo, G.D.; Bounoua, L. A revised land surface parameterization (SiB2) for atmospheric GCMs, Part I: Model formulation. *J. Clim.* **1996**, *9*, 676–705. [CrossRef]
17. Wang, L.; Koike, L.; Yang, D.; Yang, K. Improving the hydrology of the Simple Biosphere Model 2 and its evaluation within the framework of a distributed hydrological model. *Hydrol. Sci. J.* **2009**, *54*, 989–1006. [CrossRef]
18. Van Genuchten, M.T. A closed-form equation for predicting the hydraulic conductivity of unsaturated soils. *Soil Sci. Soc. Am. J.* **1980**, *44*, 892–898. [CrossRef]
19. Kuria, D.N.; Koike, T.; Lu, H.; Tsutsui, H.; Graf, H. Field-supported verification and improvement of a passive microwave surface emission model for rough, bare, and wet soil surfaces by incorporating shadowing effects. *IEEE Trans. Geosci. Remote* **2007**, *45*, 1207–1216. [CrossRef]

20. Mo, T.; Choudhury, B.J.; Schmugge, T.J.; Wang, J.R.; Jackson, T.J. A model for microwave emission from vegetation-covered fields. *J. Geophys. Res.* **1982**, *87*, 11229–11237. [CrossRef]
21. Duan, Q.Y.; Gupta, V.K.; Sorooshian, S. Shuffled complex evolution approach for effective and efficient global minimization. *J. Optim. Theory Appl.* **1993**, *76*, 501–521. [CrossRef]
22. Qin, J.; Liang, S.; Yang, K.; Kaihotsu, I.; Liu, R.; Koike, T. Simultaneous estimation of both soil moisture and model parameters using particle filtering method through the assimilation of microwave signal. *J. Geophys. Res.* **2009**, *114*, D15103. [CrossRef]
23. Bindlish, R.; Cosh, M.H.; Jackson, T.J.; Koike, T.; Fujii, H.; Chan, S.T.K.; Asanuma, J.; Berg, A.; Bosch, D.D.; Caldwell, T.; et al. GCOM-W AMSR2 Soil Moisture Product Validation Using Core Validation Sites. *IEEE J. Sel. Top. Appl. Earth Observ. Remote Sens.* **2018**, *11*, 209–219. [CrossRef]
24. Ulman, M. Diabatic heating, African Desert Locusts in Morocco in November 2004. *Brit. Birds* **2006**, *99*, 489–491.
25. Ceccato, P.; Cressman, K.; Giannini, A.; Trzaska, S. The desert locust upsurge in West Africa (2003–2005): Information on the desert locust early warning system and the prospects for seasonal climate forecasting. *Int. J. Pest Manag.* **2007**, *53*, 7–13. [CrossRef]

Article

Enhancing Ecosystem Services to Minimize Impact of Climate Variability in a Dry Tropical Forest with Vertisols

Maria Simas Guerreiro [1], Eunice Maia de Andrade [2,*], Helba Araújo de Queiroz Palácio [3], José Bandeira Brasil [4] and Jacques Carvalho Ribeiro Filho [4]

[1] Faculdade de Ciência e Tecnologia, Universidade Fernando Pessoa, Praça 9 Abril, 4249-004 Porto, Portugal; mariajoao@ufp.edu.pt

[2] Departmento de Conservação de Solo e Água, Universidade Federal Rural do Semi-Árido, Rua Francisco Mota, 572, Mossoró CEP 59625-900, Brazil

[3] Instituto Federal de Educação, Ciência e Tecnologia do Ceará, Rodovia Iguatu-Várzea Alegre, km 5, Iguatu 63503-790, Brazil; helbaraujo@ifce.edu.br

[4] Departamento de Engenharia Agrícola, Campus do Pici, Universidade Federal do Ceará, Fortaleza CEP 60455-760, Brazil; josebandeira@alu.ufc.br (J.B.B.); jacquesfilho@alu.ufc.br (J.C.R.F.)

* Correspondence: eunice.andrade@ufersa.edu.br

Abstract: Increased droughts and variable rainfall patterns may alter the capacity to provide ecosystem services, such as biomass production and clean water provision. The impact of these factors in a semi-arid region, especially on a dry tropical forest with Vertisols and under different land uses such as regenerated vegetation and thinned vegetation, is still unclear. This study analyzes hydrologic processes under precipitation pulses and intra-seasonal droughts, and suggests management practices for ecosystem services improvement. A local 43-year dataset showed a varying climate with a decrease in number of small events, and an increase in the number of dry days and in event rainfall intensity, in two catchments with different land use patterns and with Vertisols, a major soil order in semi-arid tropics. The onset of runoff depends on the expansive characteristics of the soil rather than land use, as dry spells promote micro-cracks that delay the runoff process. Forest thinning enhances groundcover development and is a better management practice for biomass production. This management practice shows a lower water yield when compared to a regenerated forest, supporting the decision of investing in forest regeneration in order to attend to an increasing water storage demand.

Keywords: semi-arid region; dry tropical forest; hydrologic processes

Citation: Guerreiro, M.S.; Maia de Andrade, E.; Palácio, H.A.d.Q.; Brasil, J.B.; Filho, J.C.R. Enhancing Ecosystem Services to Minimize Impact of Climate Variability in a Dry Tropical Forest with Vertisols. *Hydrology* **2021**, *8*, 46. https://doi.org/10.3390/hydrology8010046

Academic Editor: Philip Micklin and Pingping Luo

Received: 28 February 2021
Accepted: 14 March 2021
Published: 16 March 2021

Publisher's Note: MDPI stays neutral with regard to jurisdictional claims in published maps and institutional affiliations.

1. Introduction

Semi-arid regions, with an aridity index between 0.2 and 0.5, comprise 37% (22.6×10^6 km^2) of total dryland area, which accounts for 41% of the world's land surface [1] and half of total dryland expansion [2]. These regions are home to 14% of the world's population and their sustainability depends, besides other factors, on the scarce water availability. Human establishment in these areas demands the constant availability of water resources that compete with ecosystem maintenance requirements. Climate change scenarios suggest an increase in temperature, a reduction in rainfall and an increase in consecutive dry days, promoting an increase in dry spells and droughts [3,4].

Due to climate change, the growing season for rainfed crops is expected to decrease in length [5,6], which will imply either a reduction in yields, or a need for irrigation water to maintain productivity. Knowledge of hydrologic processes in these regions is of the utmost importance to the users and water agencies, in order to support long-term decisions in soil and water resource management to promote eco-efficiency [7,8]. The integrated management of water resources may improve the quality of living and sustain ecosystems, given the appropriate strategies and guidelines for water management.

In semi-arid tropical (SAT) regions with distinct wet and dry seasons that last for several months, we can find seasonally dry tropical forests (SDTFs) with temperatures greater than 18 °C, a potential evapotranspiration to rainfall ratio greater than 1.0 [9], summer rains and a 5- to 8-month dry period, and which are present in Africa, South America, and Asia [10]. SDTFs cover an area of approximately 10^6 km^2, with the largest extent of continuous SDTF fragments in Bolivia, Paraguay and North Argentina, and in northeast Brazil [11]. These two areas represent 54% of the world's SDTFs. A great part of these forests is degraded or is being subject to other land uses, such as cropland [12] and livestock production [13], with vegetation thinning the most common management practice for enhancing feed and biomass production.

SDTFs play a significant role in the direct provision of food, despite little research being done on the subject outside of Africa [14]. SAT regions with an average rainfall between 500 and 1000 mm have Vertisols as an important soil order [15]. There are 3.35 million km^2 of these soils in the semi-arid regions of Africa, India, China, Australia, the American southwest, and South America [16,17], particularly in the Brazilian northeast [15], of which 1.5 million km^2 are potentially agricultural areas. SDTFs are sensitive to droughts, and their resistance and resiliency to the intra- and interannual variability of rainfall need to be assessed in order to minimize impacts and adapt to challenging climate changes [9]. The dynamics of climate change in semi-arid regions might vary by location, depending on the effects of global- and regional-scale factors [18,19], hence the need to study the hydrologic processes of SDTFs.

The swelling nature of Vertisols promotes crack formation when dry, enhancing salt mobilization and salinization of the aquifer from cultivation [20], and surface sealing when wet, limiting drainage and increasing difficulty in land preparation for crop production [21]. Nonetheless, their high water-holding capacity due to the high expansive 2:1 clay content is adequate for dryland crop production in semi-arid environments under adequate management [15,21].

Some studies state that forest-to-grassland conversion in the high-elevation tropics results in little runoff increase [22], contrary to other studies that show that forest thinning promotes water yield over the native forest in temperate dry regions. The ultimate goal is to increase ecosystem services by adopting best management practices [23], such as restoring land use on marginal lands to enhance biomass production or water yield, purify surface and ground waters, and contribute to carbon sequestration [24,25]. Therefore, this study seeks to understand the watershed response to land use management in SATs under Vertisols.

The objective of this study is to propose best management practices that improve ecosystem services and minimize the effects of climate variability on SDTFs with Vertisols in semi-arid regions. To achieve this objective, we have analyzed the hydrologic response of SDTFs with expansive clay soil to variable rainfall and dry spells, and analyzed the hydrologic response to climate variability. The results may help decision makers (users and water agencies) define strategies to implement management practices that favor either water storage or biomass production in dry tropical forests.

2. Materials and Methods

2.1. Study Area

The hydrologic study was conducted on two small catchments located in the South-Central region of the state of Ceará, Brazil, part of the Federal Institute for Education, Science, and Technology of Ceará (IFCE), Campus Iguatu (Figure 1). Both catchments have an ephemeral second-order stream in areas of 2.1 ha and 1.1 ha, and average slopes of 10.6 and 8.7%, respectively.

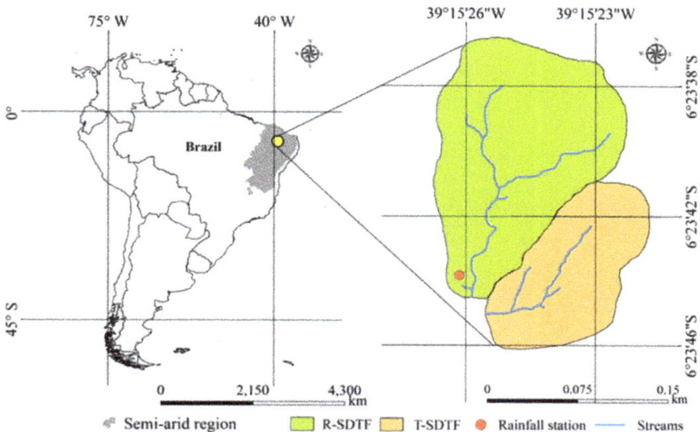

Figure 1. Catchments' locations.

Although average annual rainfall in the region is 998 mm, the annual potential evapotranspiration is 2113 mm, with a Thornthwaite's aridity index of 0.48, classifying the climate as semi-arid. This type of climate is confirmed by the Köppen climate classification (BS: semi-arid hot), and by the life zone chart introduced by Holdridge, 1968 [26], supported by the Food and Agriculture Organization [27]. This region is characterized by nine months of water deficit (Figure 2), emphasizing its semi-aridity, despite its total annual rainfall.

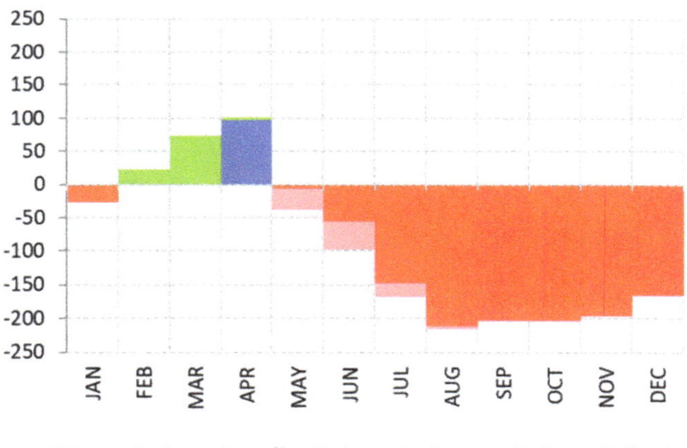

Figure 2. Average annual water balance in the period (mm).

The vegetation is typical for a seasonally dry tropical forest (SDTF) with a wide variety of spine trees in secondary succession, with a prevalence of *Aspidosperma pyrifolium*, *Commiphora leptophloeos* (Mart.) J.B. Gillett, *Mimosa caesalpiniaefolia* (Benth.), *Combretum leprosumi* (Mart.), and *Piptadenia stipulacea* (Benth.) with a height between 7 and 15 m. The two adjacent catchments show different land uses: one is a seasonally dry tropical forest under regeneration for 40 years (R-SDTF) and the other was subject to thinning (T-SDTF) in December of 2008. In the thinning management, all vegetation with a diameter below 10 cm was cut, and the branches were left on site.

The two catchments have a similar soil type, typical Calcic Vertisol (Pellic), with a high content of expansive 2:1 montmorillonite clay characterized by X-ray diffraction (XRD) and X-ray fluorescence (XRF) techniques. The soils are 2 to 3 m deep, and develop cracks when dry and are sticky and sealed when wet. Due to the high clay content, these soils show low hydraulic conductivity (Table 1), as discussed by Pathak et al. (2013) [15].

Table 1. Catchment properties for the seasonally dry tropical forest under regeneration (R-SDTF) and seasonally dry tropical forest subject to thinning (T-SDTF) [8].

Properties		R-SDTF	T-SDTF
Saturated hydraulic conductivity	m hr^{-1}	0.25	0.20
Phosphorus	g dm^{-3}	11.63	45.79
Organic matter	%	0.88	1.48
Sand	%	18.4	12.4
Silt	%	33.75	38.15
Clay	%	47.85	49.43
Soil texture		Clay	Clay
Canopy cover	%	90	60
Ground cover	%	30	100

R-SDTF: seasonally dry tropical forest under regeneration for 40 years; T-SDTF: seasonally dry tropical forest subject to thinning.

The canopy cover (90%) in the R-SDTF shades the groundcover (30%), limiting its growth (Figure 3a), whereas the sparser canopy cover (60%) in the T-SDTF allows solar radiation to reach the soil and improve the groundcover development (Figure 3b).

Figure 3. Groundcover: (**a**) R-SDTF and (**b**) T-SDTF. Photos by the author.

2.2. Data Analysis

2.2.1. Long-Term Rainfall Series

A 1974 to 2017 dataset from the Iguatu, CE, Brazil rain gauge (http://www.funceme.br, accessed on 13 May 2020) was used for the characterization of rainfall events, in the following categories: annual total, monthly distribution, monthly average continuous dry days (CDD) and continuous wet days (CWD), number of dry (DD) and wet days (WD), and mean daily rainfall. The mean annual rainfall for the studied area was 995 ± 305 mm (Figure 4). The temporal distribution of rainfall depth shows 85% to be concentrated from January to May, of which 27% occurs in March, on average.

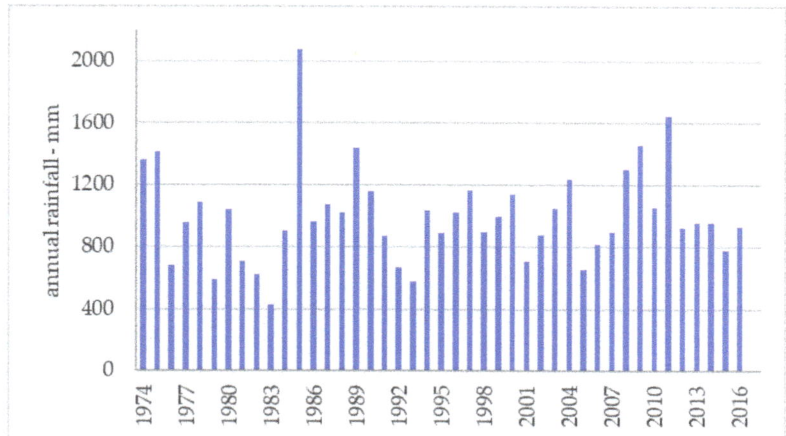

Figure 4. Long-term annual rainfall from 1974 to 2017, Iguatu, Brazil.

2.2.2. Short-Term Rainfall Runoff Events

Catchment response was analyzed based on rainfall and runoff data collected between 2009 and 2017 during the wet season (January to May) at two catchments in Iguatu, CE, Brazil. Runoff data from 2012 and 2014 were rejected due to a sensor failure. A total of 259 rainfall events were recorded at the rain gauge shared by the two adjacent catchments. The R-SDTF and T-SDTF vegetation catchments produced 67 and 60 runoff events in the period, respectively.

2.2.3. Statistics

Descriptive statistical analyses were performed on the long-term daily rainfall series for total daily rainfall, number of wet days in the year and number of wet days in the wet season (Jan to May). Histograms were developed for total annual rainfall and number of dry days in a year. Trends in annual rainfall and for annual number of dry days for the 43 year-long series were evaluated based on the significance of the slope being statistically different from zero.

Descriptive statistics were also performed on the short-term daily rainfall series associated with the runoff data in both catchments—all rainfall events, rainfall that generated runoff, and rainfall that did not generate runoff.

A *t*-test to compare water yield from both land use management strategies was performed at the 95% confidence level.

3. Results and Discussion

3.1. Long-Term Rainfall Series

The larger dataset (1974–2017) shows that the average annual rainfall in Iguatu is 995 mm, across 55 wet days. On a rainy day, the expected mean value is 18.1 mm, and the median is 11.0 mm (Table 2).

Based on the temporal variability of rainfall events in semi-arid regions [28], total annual rainfall and number of dry days were used to define a dry year (25th percentile) and a wet year (75th percentile), with a total annual rainfall below 790 mm and more than 323 dry days, and a rainfall above 1128 mm and less than 302 dry days, respectively (Figure 5b and Table 2).

Table 2. Characteristics of daily rainfall at Iguatu, CE, Brazil (1974–2017).

		Daily Rainfall (mm)	Number of Wet Days	
			Annual	Jan–May
N (number)		16014	2418	2109
Mean		2.7	18.1	18.4
Std. Deviation		10.0	19.7	19.7
Minimum		0.00	0.20	0.20
Maximum		174.0	174.0	174.0
Percentiles	25th	0.0	5.0	5.0
	50th	0.0	11.0	11.0
	75th	0.0	24.0	25.0

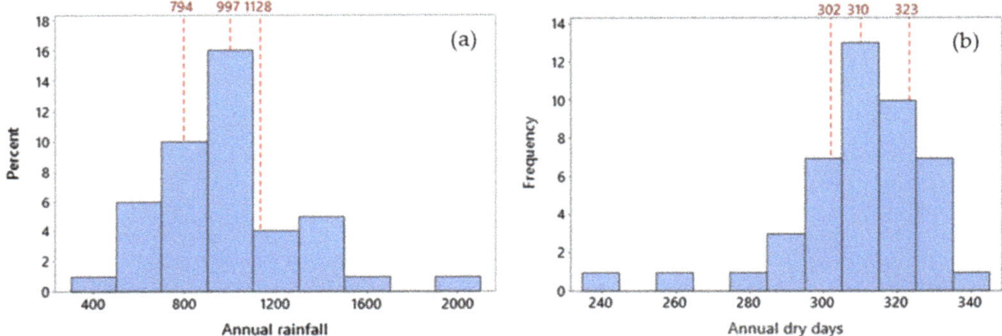

Figure 5. Histogram of (**a**) annual rainfall and (**b**) number of dry days (DD) (1974–2017).

Even though total annual rainfall did not show a significant trend in the period (1974–2017), the number of dry days and average rainfall intensity increased significantly (p-value = 0.05 and 0.001, respectively). These results suggest that extreme events are more frequent (dry spells and intense rainfall events).

The rain falls mostly between December and May, and seasonal dry spells occur from June to December (Figure 6a), as continuous dry days (CDD) increase from an average of 21 in July to almost 70 in November (Figure 6c). Total monthly rainfall shows a higher skewness from June to November, the mean value being a poor parameter to represent this variable. The average number of CDD from January to May is less than a week, and continuous wet days (CWD) tend to occur in pairs (1.5 to 2.5 days) in the wet season (January to May) and one at a time in the dry season (June to December), in small amounts (Figure 6b).

The rainfall driving forces are cold fronts from October to January, which displace the subtropical cells in the Atlantic Ocean and lead to an increase in monthly precipitation. The intertropical convergence zone (ITCZ) from January to May, which reaches the highest southern latitude (6° S) in March, sets the rainfall season, after which it goes back to the northern hemisphere, allowing the easterly waves to attenuate rainfall occurrence from May to August [28]. Subtropical cells in the Atlantic Ocean are responsible for the driest season from August to October [29], closing the annual cycle.

3.2. Short Term Rainfall-Runoff Events

The catchment dataset (2009–2017) shows that the average annual rainfall in the wet season is 759 mm (507 mm to 1308 mm), over 37 (26 to 54) wet days. On a rainy day, the expected mean value is 20.4 mm, and the median is 14.3 mm, of which a minimum of 5.3 mm is generated runoff (Table 3).

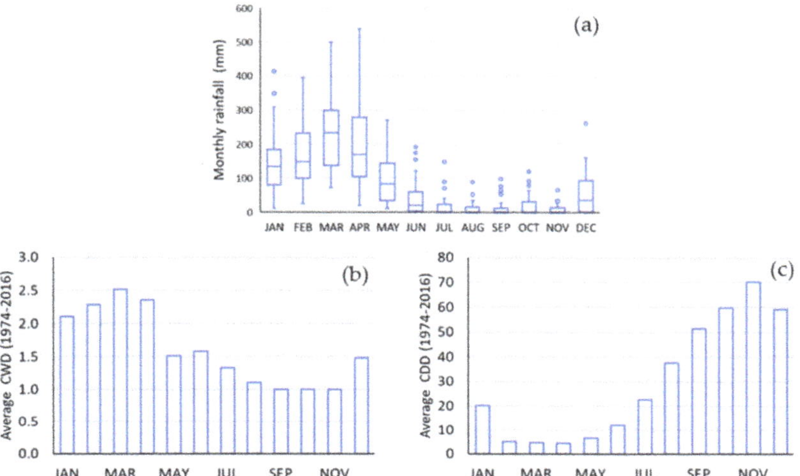

Figure 6. Monthly (**a**) rainfall; (**b**) continuous wet days (CWD); (**c**) continuous dry days (CDD)—1974–2017.

Table 3. Characteristics of rainfall and runoff at catchment scale (2009–2017).

		Daily Rainfall Event (mm)			
		All Events	Runoff Generating	No Runoff Generating	Runoff
N (number)		257	67	190	67
Mean		20.4	34.9	15.3	5.4
Std. Deviation		19.3	23.7	14.4	7.2
Minimum		1.3	5.3	1.3	0.01
Maximum		162.0	162.0	80.3	34.5
Percentiles	25th	7.2	19.8	6.1	0.6
	50th	14.3	30.3	11.4	2.6
	75th	27.9	47.9	18.2	7.0

The number of rainfall events during the wet season shows little relationship with total rainfall (Table 4), as discussed by Guerreiro et al. (2013) [28]—the second year with the most events (46) was the second driest year of the series, 2017. In the period for which there is runoff data (2009–2017), 2015 and 2017 were classified as dry years and 2009 and 2011 as wet years.

The amount of rainfall prior to the first runoff event of the water year (P_{cum} to Q1) is not proportional to the number of events (Table 4) but has a positive relationship with the number of days from first rainfall to first runoff event, ΔT. The ratio between P_{cum} to Q1 and ΔT is a function of the soil type, the expansive Vertisol [30–32]. The seasonal dry spell from June to December promotes wide cracks due to the shrinkage of the Vertisol at the end of the dry season, which influences infiltration and the resulting runoff at the beginning of the wet season [15]. The dry spells during the wet season promote micro-cracks, and delay the runoff process, as will be explained in the following section.

Table 4. Rainfall characteristics: rainfall runoff data set (2009–2017).

	P_{cum} (mm)	# P_{events}	P_{cum} to Q1 (mm)		ΔT (days)		Q_{total} (mm)		# Q_{events}		P_{min} to Q (mm)		P_{max} to no Q (mm)
Year	Wet Season		R	T	R	T	R	T	R	T	R	T	
2009	981.0	39	162.0	162.0	8	8	104.0	74.0	25	26	8.6	9.0	68
2010	711.0	26	163.0	163.0	23	23	15.0	11.0	7	9	28.7	10.0	80
2011	1308.0	54	182.0	222.0	21	25	188.0	143.0	26	22	12.0	12.0	69
2012	SENSOR FAILURE												
2013	634.0	28	475.0	475.0	126	126	39.0	30.0	2	2	44.0	44.0	59
2014	SENSOR FAILURE												
2015	507.0	32	113.0	0.0	52		0.1	0.0	1	0	5.3		56
2016	608.0	34	345.0	345.0	80	80	7.0	1.0	2	1	8.0	12.0	66
2017	583.0	46	327.0	0.0	49		8.0	0.0	4	0	2.0		40

P_{cum}—cumulative rainfall; # P_{events}—number of rainfall events; P_{cum} to Q1—cumulative rainfall to first runoff event; ΔT—number of days from first rainfall to first runoff event; Q_{total}—total annual runoff; # Q_{events}—total number of runoff events; P_{min} to Q—minimum daily rainfall to promote a runoff event; P_{max} to no Q—maximum daily rainfall that did not produce runoff; R (Regenerated dry tropical forest catchment); T (Thinned dry tropical forest catchment).

A greater ΔT in dry years lead to a loss of soil moisture, the shrinkage of expansive clays [30], and the presence of micro-cracks [31], demanding more rainfall to start the runoff process [32]. Cracks in Vertisols increase hydraulic conductivity [33], promoting preferential flow paths for water infiltration in the soil. Even after the sealing of the cracks, there might still be preferential flow paths [34], explaining the cumulative rainfall requirement of up to 475 mm for runoff to occur.

Even though the two land uses under study, R-SDTF and T-SDTF, showed similar responses to cumulative rainfall and time to first runoff event, a *t*-test showed a different response to water yield (*p*-value $\leq$ 0.05). The similarities among the two catchments regarding the onset of the first runoff event suggest that the beginning of runoff depends on the characteristics of the Vertisol soil, and the differences suggest a dependency on ground-cover (Figure 3), since the two catchments have similar geomorphological characteristics and soil type, but different land uses (Table 1).

The R-SDTF catchment had runoff events in all years of study (2009–2017), whereas the T-SDTF did not exhibit runoff during the dry years 2015 and 2017 (Figure 7b). The cracks generated in the soil prior to the beginning of the wet season increase initial abstractions, influence infiltration, and affect the onset of runoff generation, as suggested by [15]. The onset of runoff in both catchments needs a minimum cumulative rainfall of 62 mm within a 7-day period (PCUM 7) from the previous rainfall event (Figure 7a,b).

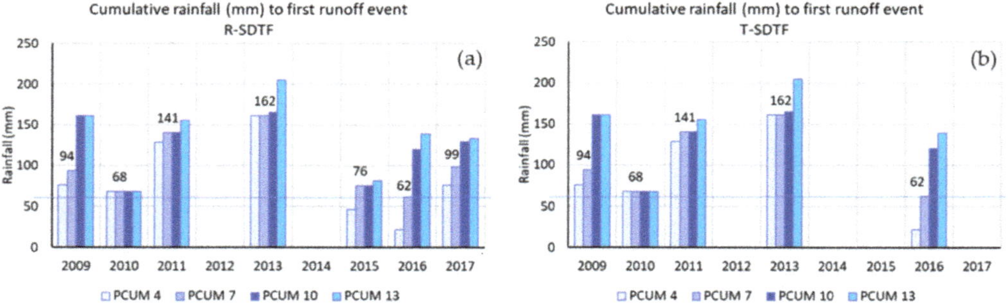

Figure 7. Rainfall characteristics to first runoff event (2009-2017): (**a**) R-SDTF; (**b**) T-SDTF.

T-SDTF reacted to thinning by increasing the number of runoff events but decreasing total yield. The increase in number of events was expected, but not the decrease in water yield [23,35]. This result suggests that the processes were still not established in SDTFs after thinning. However, after 2011 (two years after forest thinning), the groundcover had already been improved and both the number of events and water yield showed a decrease (Figure 7b), as already verified by Ponette-González et al. (2013) [22]. These results suggest that groundcover development reduces surface flow due to an increase in surface roughness that reduces flow velocity and increases infiltration opportunity [8], and the development of a herbaceous root system [36] that creates preferential flow paths [30] for water to infiltrate the soil [22].

Total annual rainfall from the time series (1974–2017) classified 2009 and 2011 as wet years, and 2015 and 2017 as dry years. Dry years (annual rainfall < 608 mm) showed low to no runoff, due to the shrinking and swelling characteristics of the soils [15]. During the dry years 2015 and 2017, despite the cumulative rainfall being above 62 mm (Figure 7b), the low continuous dry days (CDD) (lower than seven) associated with rainfall events below 40 mm (Figure 8a,b) promoted and maintained groundcover in the thinned vegetation catchment, increasing roughness and infiltration opportunity, with no resulting runoff.

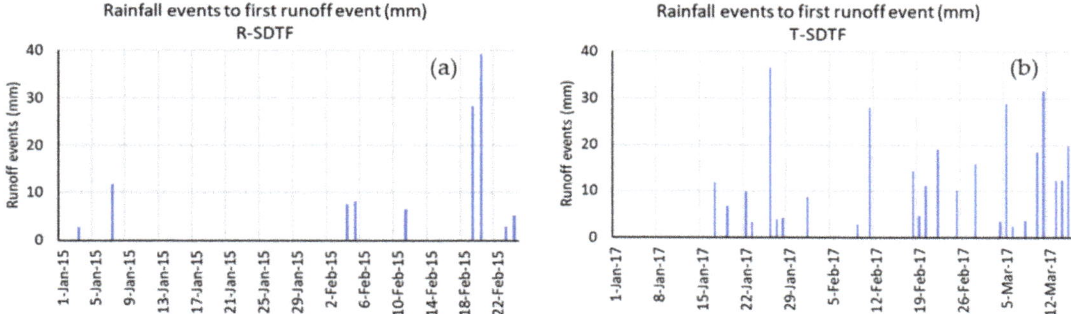

Figure 8. Rainfall events leading to first runoff event in the dry years 2015 (**a**) and 2017 (**b**).

The catchments responded similarly to the onset of runoff when there was enough CDD to compromise the development of vegetation in the T-SDTF, mimicking the R-SDTF catchment, suggesting that the soil is the major driver of the process. When the CDD is above 30 days, micro-cracks are formed, and the soil returns to the initial dry, cracked condition. The median value of cumulative rainfall to the beginning of the runoff process is approximately 200 mm when the maximum CDD is below 30 days [37]. This was particularly evident in 2016, when a total cumulative rainfall of 345 mm was necessary to begin the runoff process, after a dry spell of 36 days in which a total rainfall of 298 mm occurred.

The behavior contributing to the onset of runoff events is evident for the wet years 2009 to 2011 and the average years 2013 and 2016, but none can be spotted for the dry years 2015 and 2017 (Figure 9a). In average and wet years, there is a linear relationship between cumulative rainfall for the onset of the first runoff event and CDD. More CDD in dry years results in more water loss via evapotranspiration and soil moisture reduction, and reinstates cracks in the soil [34], which redefine new preferential flow paths for water to infiltrate into the soil. Still, Figure 9b shows a similar behavior in both catchments, confirming that soil, rather than land use, is the major driver to start the runoff process in Vertisols in a semi-arid tropical region.

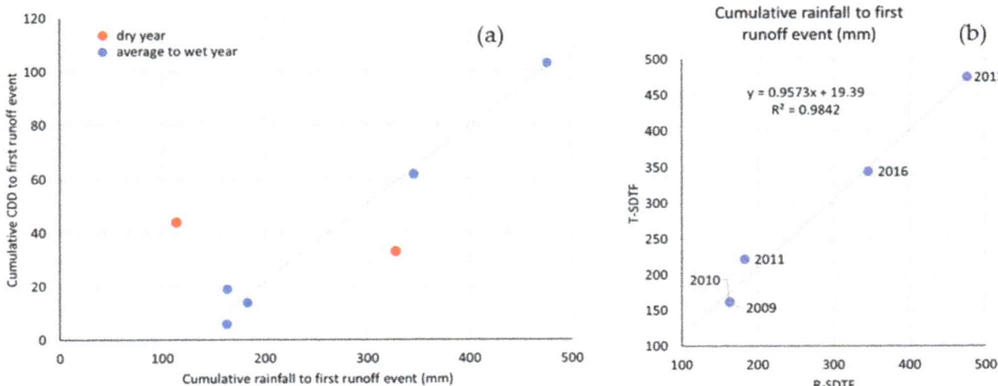

Figure 9. (a) CDD vs. rainfall to first runoff event; **(b)** rainfall to first runoff event: regenerated vs. thinned vegetation

Once runoff has been established, groundcover explains the variation in water yield in both catchments, as the water yield from the T-SDTF catchment corresponds to approximately 75% of the R-SDTF (Figure 10). This suggests that 25% of total rainfall will be available to the other eco-hydrological processes. T-SDTF allows solar radiation to hit the soil and promotes the development of groundcover [36,38], which increases surface roughness and infiltration opportunity and reduces runoff. Regenerated vegetation promotes greater coverage of the soil, but canopy shading does not allow the further development of groundcover. Biomass production may be a profitable alternative to agriculture, as crop yield is expected to be more affected by climate variability than perennial biomass [39].

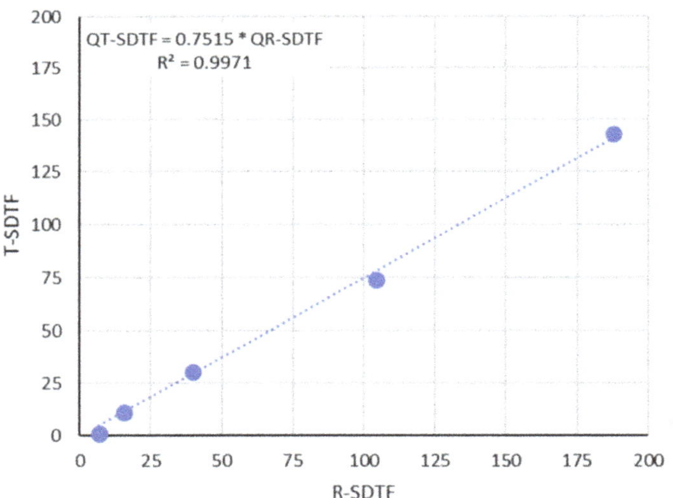

Figure 10. T-SDTF vs. R-SDTF water yield relationship.

4. Conclusions

Forest thinning enhances groundcover development and is a better management practice for biomass production. This management practice shows a lower water yield when compared to a regenerated forest, supporting the decision of investing in forest regeneration to attend to an increasing water storage demand. The number of dry days and rainfall intensity increased in the 1974–2017 period, aggravating the already dry characteristics of the region, and contributing to more extreme runoff events. In the case of a dry spell over 30 days, micro-cracks are formed, and the soil returns to the initial dry, cracked condition. The onset of runoff events is defined by the soil characteristics rather than the land use. The onset of runoff in both catchments requires a minimum cumulative rainfall of 62 mm within a 7-day period from the previous rainfall event, although it is the land-use that defines water yield. The impact of water storage in the soil cracks upon the onset of runoff generation is evident, but needs further studies for quantification.

Author Contributions: All authors made a significant contribution to the final version of the manuscript. Conceptualization, M.S.G., E.M.d.A. and H.A.d.Q.P.; methodology, M.S.G., J.B.B. and J.C.R.F.; writing—original draft preparation, M.S.G., E.M.d.A. and H.A.d.Q.P.; writing—review and editing, all authors; supervision, M.S.G. and E.M.d.A.; project administration, E.M.d.A. All authors have read and agreed to the published version of the manuscript.

Funding: This work was supported by CNPq—Conselho Nacional de Desenvolvimento Científico e Tecnológico, Brazil [grant number 558135/2009-9].

Informed Consent Statement: Not applicable.

Data Availability Statement: The data presented in this study are available on request from the corresponding author.

Acknowledgments: This study was carried out with the support of the Coordenação de Aperfeiçoamento de Pessoal de Nível Superior - Brasil (CAPES), the Conselho Nacional de Desenvolvimento Científico e Tecnológico (CNPq) and the Fundação Cearense de Apoio ao Desenvolvimento Científico e Tecnológico.

Conflicts of Interest: The authors declare that they have no conflict of interest.

References

1. UN. 2010–2020: UN Decade for Deserts and the Fight Against Desertification. [WWW Document]. 2019. Available online: https://www.un.org/en/events/desertification_decade/whynow.shtml (accessed on 13 May 2020).
2. Huang, J.; Ji, M.; Xie, Y. Global semi-arid climate change over last 60 years. *Clim. Dyn.* **2016**, *46*, 1131–1150. [CrossRef]
3. Huang, J.; Li, Y.; Fu, C.; Chen, F.; Fu, Q.; Dai, A.; Shinoda, M.; Ma, Z.; Guo, W.; Li, Z.; et al. Dryland climate change: Recent progress and challenge. *Rev. Geophys.* **2017**, 719–778. [CrossRef]
4. Marengo, J.A.; Torres, R.R.; Alves, L.M. Drought in Northeast Brazil—Past, present, and future. *Theor. Appl. Clim.* **2016**, *129*, 1189–1200. [CrossRef]
5. Kihupi, N.I.; Tarimo, A.K.P.R.; Masika, R.J.; Boman, B.; Dick, W.A. Trend of Growing Season Characteristics of Semi-Arid Arusha District in Tanzania. *J. Agric. Sci.* **2015**, *7*, 45–55. [CrossRef]
6. Gebru, T.A.; Brhane, G.K.; Gebremedhin, Y.G. Contributions of water harvesting technologies intervention in arid and semi-arid regions of Ethiopia, in ensuring households' food security, Tigray in focus. *J. Arid Environ.* **2021**, *185*, 104373. [CrossRef]
7. Lal, R. Sequestering carbon and increasing productivity by conservation agriculture. *J. Soil Water Conserv.* **2015**, *70*, 55–62. [CrossRef]
8. Andrade, E.M.; Guerreiro, M.J.S.; Palácio, H.A.Q.; Campos, D.A. Ecohydrology in a Brazilian tropical dry forest: Thinned vegetation impact on hydrological functions and ecosystem services. *J. Hydrol. Reg. Stud.* **2020**, *27*, 100649. [CrossRef]
9. Allen, K.; Dupuy, J.M.; Gei, M.G.; Hulshof, C.; Medvigy, D.; Pizano, C.; Salgado-Negret, B.; Smith, C.M.; Trierweiler, A.; Bloem, S.J.; et al. Will seasonally dry tropical forests be sensitive or resistant to future changes in rainfall regimes? *Environ. Res. Lett.* **2017**, *12*, 1–16. [CrossRef]
10. FAO. Global Ecological Zoning for the Global Forest Resources Assessment 2000—Final Report. Rome. [WWW Document]. 2001. Available online: http://www.fao.org/3/ad652e/ad652e00.htm (accessed on 13 May 2020).
11. Miles, L.; Newton, A.C.; DeFries, R.S.; Ravilious, C.; May, I.; Blyth, S.; Kapos, V.; Gordon, J.E. A global overview of the conservation status of tropical dry forests. *J. Biogeogr.* **2006**, *33*, 491–505. [CrossRef]
12. Alenius, T.; Mökkönen, T.; Holmqvist, E.; Ojala, A. Neolithic land use in the northern Boreal zone: High-resolution multiproxy analyses from Lake Huhdasjärvi, south-eastern Finland. *Veg. Hist. Archaeobotany* **2017**, *26*, 469–486. [CrossRef]
13. Stuhler, J.; Orrock, J. Historical land use and present-day canopy thinning differentially affect the distribution and abundance of invasive and native ant species. *Biol. Invasions* **2016**, *18*, 813–1825. [CrossRef]
14. Blackie, R.; Baldauf, C.; Gautier, D.; Gumbo, D.; Kassa, H.; Parthasarathy, N.; Paumgarten, F.; Sola, P.; Pulla, S.; Waeber, P.; et al. *Tropical Dry Forests the State of Global Knowledge and Recommendations for Future Research*; Cifor: Bogor, Indonesia, 2014; Volume 2.
15. Pathak, P.; Sudi, R.; Wani, S.P.; Sahrawat, K.L. Hydrological behavior of Alfisols and Vertisols in the semi-arid zone: Implications for soil and water management. *Agric. Water Manag.* **2013**, *118*, 12–21. [CrossRef]
16. Driessen, P.; Deckers, J.; Spaargaren, O.; Nachtergaele, F. *Lecture Notes on the Major Soils of the World*; Food and Agriculture Organization (FAO): Rome, Italy, 2000.

17. Latham and Peter Ahn. Management of Vertisols under Semi-Arid Conditions, IBSRAM PROCEEDINGS No. 6. 1986. Available online: https://horizon.documentation.ird.fr/exl-doc/pleins_textes/divers15-08/010065123.pdf (accessed on 13 May 2020).
18. Zuo, D.; Xu, Z.; Yao, W.; Jin, S.; Xiao, P.; Ran, D. Assessing the effects of changes in land use and climate on runoff and sediment yields from a watershed in the Loess Plateau of China. *Sci. Total Environ.* **2016**, *544*, 238–250. [CrossRef]
19. Yair, A. Contrasting hydrological regimes in two adjoining semi-arid areas, with low rain intensities. *J. Arid Environ.* **2021**, *187*, 104404. [CrossRef]
20. Kurtzman, D.; Baram, S.; Dahan, O. Soil-aquifer phenomena affecting groundwater under vertisols: A review. *Hydrol. Earth Syst. Sci.* **2016**, *20*, 1–12. [CrossRef]
21. Debele, T.; Deressa, H. Integrated Management of Vertisols for Crop Production in Ethiopia: A Integrated Management of Vertisols for Crop Production in Ethiopia: A Review. *J. Biol. Agric. Healthc.* **2016**, *6*, 26–36.
22. Ponette-González, A.; Marín-Spiotta, E.; Brauman, K.; Farley, K.; KC, K.W.; Young, K. Hydrologic Connectivity in the High-Elevation Tropics: Heterogeneous Responses to Land Change. *Bioscience* **2013**, *64*, 92–104. [CrossRef]
23. Wu, Y.; Zhang, X.; Li, C.; Xu, Y.; Hao, F.; Yin, G. Ecosystem service trade-offs and synergies under influence of climate and land cover change in an afforested semiarid basin, China. *Ecol. Eng.* **2021**, *159*, 106083. [CrossRef]
24. Lal, R. Saving global land resources by enhancing eco-efficiency of agroecosystems. *J. Soil Water Conserv.* **2018**, *73*, 100A–106A. [CrossRef]
25. Lal, R. Soil carbon sequestration to mitigate climate change. *Geoderma* **2004**, *123*, 1–22. [CrossRef]
26. Holdridge, L.R. Life Zone Ecology. *For. Sci.* **1968**, *14*, 382. [CrossRef]
27. FAO. Global Ecological Zones for Fao Forest Reporting: 2010 Update. Rome. 2012. Available online: http://www.fao.org/3/ap861e/ap861e00.pdf (accessed on 13 May 2020).
28. Guerreiro, M.J.S.; Andrade, E.M.; Abreu, I.; Lajinha, T. Long-term variation of precipitation indices in Ceará State, Northeast Brazil. *Int. J. Climatol.* **2013**, *33*, 2929–2939. [CrossRef]
29. Moron, V.; Ward, M.; Navarra, A. Observed and SST-Forced seasonal Rainfall Variability Across Tropical America. *Int. J. Climatol.* **2001**. [CrossRef]
30. Dinka, T.M.; Morgan, C.L.S.; Mcinnes, K.J.; Kishné, A.S.; Harmel, R.D. Shrink-swell behavior of soil across a Vertisol catena. *J. Hydrol.* **2013**, *476*, 352–359. [CrossRef]
31. Li, J.H.; Li, L.; Chen, R.; Li, D.Q. Cracking and vertical preferential fl ow through land fi ll clay liners. *Eng. Geol.* **2016**, *206*, 33–41. [CrossRef]
32. Santos, J.C.N.; Andrade, E.M.; Guerreiro, M.J.S.; Medeiros, P.H.A.; Queiroz Palácio, H.A.; Araújo Neto, J.R. Effect of dry spells and soil cracking on runoff generation in a semiarid micro watershed under land use change. *J. Hydrol.* **2016**, *541*. [CrossRef]
33. Krisnanto, S.; Rahardjo, H.; Fredlund, D.G.; Leong, E.C. Water content of soil matrix during lateral water flow through cracked soil. *Eng. Geol.* **2016**, *210*, 168–179. [CrossRef]
34. Greve, A.; Andersen, M.; Acworth, R. Investigations of soil cracking and preferential flow in a weighing lysimeter filled with cracking clay soil. *J. Hydrol.* **2010**, *393*, 105–113. [CrossRef]
35. Ma, J.; Kang, F.; Cheng, X.; Han, H. Moderate thinning increases soil organic carbon in Larix principis-rupprechtii (Pinaceae) plantations. *Geoderma* **2018**, *329*, 118–128. [CrossRef]
36. Hata, K.; Kawakami, K.; Kachi, N. Higher Soil Water Availability after Removal of a Dominant, Nonnative Tree (Casuarina equisetifolia Forst.) from a Subtropical Forest. *Pac. Sci.* **2015**, *69*, 445–460. [CrossRef]
37. Greve, A.; Acworth, R.; Kelly, B. Detection of subsurface soil cracks by vertical anisotropy profiles of apparent electrical resistivity. *Geophysics* **2010**, *75*, 85–93. [CrossRef]
38. Asaye, Z.; Zewdie, S. Fine root dynamics and soil carbon accretion under thinned and un-thinned Cupressus lusitanica stands in, Southern Ethiopia. *Plant Soil* **2013**, *366*, 261–271. [CrossRef]
39. Regan, C.M.; Connor, J.D.; Segaran, R.R.; Meyer, W.S.; Bryan, B.A.; Ostendorf, B. Climate change and the economics of biomass energy feedstocks in semi-arid agricultural landscapes: A spatially explicit real options analysis. *J. Environ. Manag.* **2017**, *192*, 171–183. [CrossRef] [PubMed]

Article

Alteration of the Ecohydrological Status of the Intermittent Flow Rivers and Ephemeral Streams due to the Climate Change Impact (Case Study: Tsiknias River)

Soumaya Nabih [1], Ourania Tzoraki [2,*], Prodromos Zanis [3], Thanos Tsikerdekis [4], Dimitris Akritidis [3], Ioannis Kontogeorgos [2] and Lahcen Benaabidate [1]

[1] Laboratory of Functional Ecology and Environment Engineering, Department of Environment, Faculty of Sciences and Techniques, University of Sidi Mohamed Ben Abdellah, 30000 Fez, Morocco; soumaya.nabih@usmba.ac.ma (S.N.); Lahcen.benaabidate@usmba.ac.ma (L.B.)

[2] Department of Marine Sciences, School of the Environment, University of the Aegean, 81100 Mytilene, Greece; ikontogeorgos@isc.tuc.gr

[3] Department of Meteorology and Climatology, University Campus, Aristotle University of Thessaloniki, 54124 Thessaloniki, Greece; zanis@geo.auth.gr (P.Z.); dakritid@geo.auth.gr (D.A.)

[4] SRON Netherlands Institute for Space Research, 3584 CA Utrecht, The Netherlands; tsike@geo.auth.gr

* Correspondence: rania.tzoraki@aegean.gr

Abstract: Climate change projections predict the increase of no-rain periods and storm intensity resulting in high hydrologic alteration of the Mediterranean rivers. Intermittent flow Rivers and Ephemeral Streams (IRES) are particularly vulnerable to spatiotemporal variation of climate variables, land use changes and other anthropogenic factors. In this work, the impact of climate change on the aquatic state of IRES is assessed by the combination of the hydrological model Soil and Water Assessment Tool (SWAT) and the Temporary Rivers Ecological and Hydrological Status (TREHS) tool under two different Representative Concentration Pathways (RCP 4.5 and RCP 8.5) using CORDEX model simulations. A significant decrease of 20–40% of the annual flow of the examined river (Tsiknias River, Greece) is predicted during the next 100 years with an increase in the frequency of extreme flood events as captured with almost all Regional Climate Models (RCMs) simulations. The occurrence patterns of hyporheic and edaphic aquatic states show a temporal extension of these states through the whole year due to the elongation of the dry period. A shift to the Intermittent-Pools regime type shows dominance according to numerous climate change scenarios, harming, as a consequence, both the ecological system and the social-economic one.

Keywords: hydrologic modeling; SWAT; climate change; intermittent flow; aquatic states; TREHS tool; CORDEX model; IRES; Tsiknias River

Citation: Nabih, S.; Tzoraki, O.; Zanis, P.; Tsikerdekis, T.; Akritidis, D.; Kontogeorgos, I.; Benaabidate, L. Alteration of the Ecohydrological Status of the Intermittent Flow Rivers and Ephemeral Streams due to the Climate Change Impact (Case Study: Tsiknias River). *Hydrology* **2021**, *8*, 43. https://doi.org/10.3390/hydrology8010043

Academic Editor: Philip Micklin

Received: 29 January 2021
Accepted: 1 March 2021
Published: 5 March 2021

Publisher's Note: MDPI stays neutral with regard to jurisdictional claims in published maps and institutional affiliations.

1. Introduction

The water framework directive (WFD) established an integrated approach on management and protection of Europe's aquatic environment and set the general goals to achieve a "good water status" for European water bodies [1]. These goals, however, seemed more directed towards permanent rivers, neglecting the important contribution of intermittent streams; these are defined as all temporary, ephemeral, seasonal, and episodic streams and rivers in defined channels, in which flow is interrupted either spatially or in time [2]. Ecosystem services provided by IRES are strongly affected by their hydrological phases, with dry phase having potentially negative impact on several services and especially agricultural production [3].

Temporary streams constitute more than 50% of the global network, and this number is growing due to climate change; the status of the majority of rivers is switching from perennial to intermittent [4]. According to the Fifth Assessment Report (AR5) of the Intergovernmental Panel on Climate Change (IPCC), surface temperature is projected to rise over the 21 century under all assessed emission scenarios, while precipitation trends will not be uniform over the Earth's different regions [5]. It is very likely that heat waves will occur more often and last longer and that extreme precipitation events will become more intense and frequent, particularly in the Mediterranean area [5–7]. For instance, large-scale predictions for Mediterranean suggest up to 35% rainfall reduction and 3–5 °C temperature increase by 2071–2100 [8,9]. A climate change impact study conducted at basin scale in Portugal under RCP 4.5 and RCP 8.5 scenarios, comparing past (1950–2015) and future (2021–2100) climate, reached some concerning conclusions: annual temperatures are expected to increase by 10–20%, precipitation will decrease by 8–13%, and river flow will decrease by 28% [10]. Furthermore, based on the international Coordinated Regional Downscaling Experiment (CORDEX) simulations, the CORDEX ensemble corroborates the fact that the Mediterranean is already entering the 1.5 °C climate warming era. The southern part of the Mediterranean is expected to be impacted most strongly since the CORDEX ensemble suggests substantial combined warming and drying, particularly for pathways RCP 4.5 and RCP 8.5 [11]. Climate change will exacerbate the problems of water scarcity that will be more pronounced during the dry season; the transition from perennial to intermittent status for many rivers due to CC coupled with the effect of anthropogenic pressures on water resources will affect their ecological status [12,13].

All the above indicate the importance of the hydrological status of rivers and temporary streams in particular. The latter are less integrated in regional and global analysis, water legislation, and regulations because of the difficulties they display when compared to permanent rivers [14], especially in the ecological aspect of the assessment. The shifting lotic, lentic, and terrestrial habitats are unique in spatial arrangement and connectivity. They are controlled mainly by the magnitude, frequency, and duration of drying spells in these systems [4]. The increase, however, in fragmentation of rivers networks results in dispersal-limited freshwater ecosystems influencing the metacommunity dynamics [15]. In this perspective, the biodiversity and several ecosystem services are vulnerable to the river intermittency especially under climate change. These habitats are threatened with extinction by the projected climate change impacts.

In recent years, several initiatives and projects have been launched especially in the Mediterranean region, resulting in the introduction of new tools such as the TREHS tool for temporary rivers [16]. No studies have been conducted thus far using the TREHS tool (outside of the EU LIFE TRIVERS project), utilizing its capacity to assess the hydrological regime of temporary rivers. TREHS is used to classify the intermittent streams regime and status/degree of alteration based on metrics from hydrologically related data. The latter are usually limited or absent since most temporary streams are not monitored; the use, however, of hydrological models such as SWAT can help overcome such limitations [17]. SWAT is a widely used, physically-based, semi-distributed model which simulates the hydrological regime, generating a longer flow time series based on readily available meteorological data [18,19]. Limited efforts, however, have been put into the assessment of the hydrological regime of temporary streams in the Mediterranean region, considering climate change challenges, coupling both hydrological modeling and ecological status tools [12,20–25].

The case study is Tsiknias River (Greece), a typical intermittent flow stream, with a mainly agricultural basin, subject to severe floods and drought phenomena [26], outflowing into a protected conservation area [27]. The limited existing monitoring of the basin in combination with the interseasonal flow variability make the use of hydrological models essential to simulate its hydrologic response in longer periods. The basin adaptation to the climate change is important to maintain the current agricultural uses such as olive groves and vegetables production and conserve land productivity in the future or shift to deficit irrigation or other water stress tolerant plants.

In light of the above, the objectives of the current paper are:

- The development and comparison of historical and future hydrological simulations by means of climatic datasets generated by multi-model ensembles of RCMs, under two different greenhouse gas emission scenarios RCP 4.5 and 8.5.
- The assessment of the potential effect of climate change, under different scenarios, on the hydrologic regime of a Mediterranean intermittent river basin and the analysis of temporal streamflow trends.
- The investigation of the transition of the different aquatic states of the stream especially from flood to edaphic, crucial for the sustainability of the ecosystem biodiversity, using the TREHS tool.

2. Study Area and Datasets

2.1. Study Area

Tsiknias is an intermittent Mediterranean stream located in the central part of Lesvos Island (Figure 1). It originates from Lepetymnos Mountain (968 m) with a mainly north to south direction and it has one of the largest drainage networks on the island [28]. The Gulf of Kalloni where this stream discharges is considered as the main ecological and touristic asset of Lesvos Island. Due to its importance regarding biodiversity and the presence of endemic species, it was established as a Natura 2000 Special Conservation Area [29].

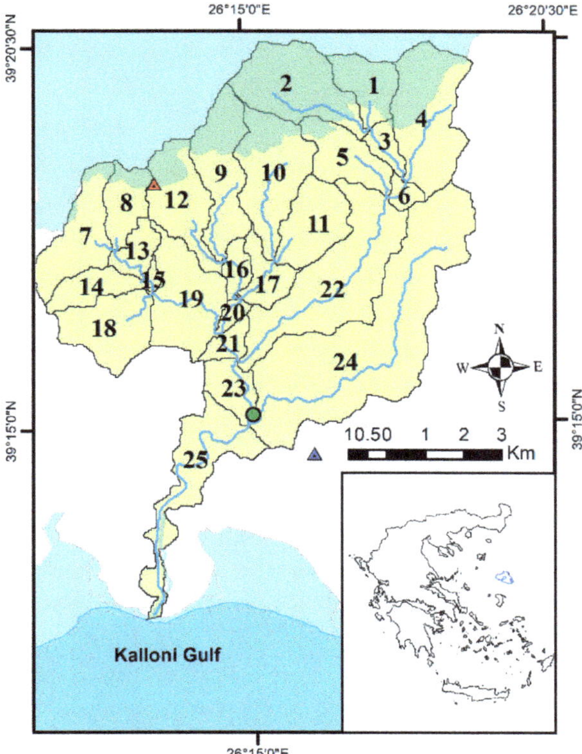

Figure 1. Map of the study area with stream network. The numbers represent subbasin division from SWAT model. Triangles represent meteorological stations: red for Stypsi station and blue for Agia Paraskevi. The circle represents Prini gauging station. The light blue represents Natura 2000 areas of the island.

Tsiknias basin is located between approximately 26°19′17.8026″ east and 26°10′52.1754″ west longitude, and between 39°11′50.7546″ south and 39°20′39.4902″ north latitude covering an area of approximately 90 km^2. The main geologic formations are of Neogene and Miocene age volcanic acidic rocks including ignimbrite, basalt, lavas, and tuffs and Pleistocene and Holocene continental deposits in the coastal part [30,31]. Most soils in this basin result from these formations. The permeability of the soils is low in most of the area, low-medium in the highland areas, and medium in the northern part of the area [32]. The major land use is agriculture (58.3%, of which olive groves cover 22.9%), followed by ryegrass (21.3%), and the remaining 20.4% is rangeland, forests, wetlands, and low-density urban areas [33].

The basin has a typical Mediterranean climate, with warm, dry summers and mild, moderately rainy winters. The mean annual temperature is 19.2 °C, and the mean annual rainfall varies from 600 mm on the plains to over 900 mm in the mountains. Since 1955, temperature has shown a slight non-uniform warming trend, while precipitation is indicating a slight decrease (Figure 2).

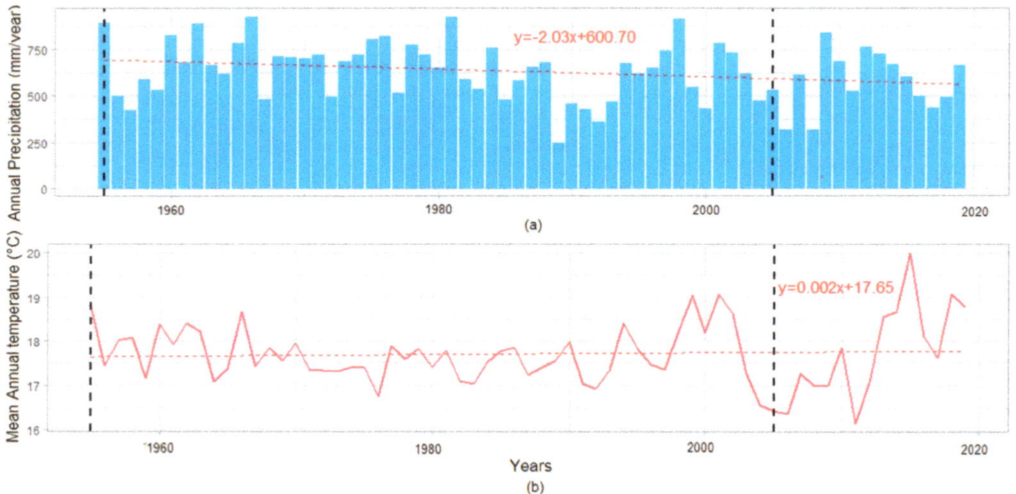

Figure 2. (**a**) Annual rainfall over Tsiknias basin from Agia Paraskevi station (95 m); and (**b**) mean annual temperature in Tsiknias basin from Agia Paraskevi station. Vertical black dashed lines delimit the period of 1955–2005, while the horizontal dashed red line corresponds to the trend.

2.2. Spatiotemporal Datasets

Table 1 includes the datasets used in this study, including the Land-Use Land-Cover (LULC), Digital Elevation Model (DEM), soil datasets, meteorological forcing datasets, and observed streamflow in the selected basin. Rainfall data were obtained from rain gauge observations at both Agia Paraskevi (95 m) and Stypsi (396 m) stations. Since 2014, a telemetric station (Prini) has been operating on the main channel, providing water level data in 15 min intervals. Monthly flow measurements from the same area are used to produce the rating curve, which relates the water level to the flow.

Table 1. Datasets used in the present study.

Dataset	Source	Frequency	Time Period	Remarks
Rainfall	Automatic meteorological station at Agia Paraskevi	Daily	(01/01/1955– 01/01/2020)	Data gaps filled with Inverse Distance Method (IDM) using data from Mytilini airport as reference station.
Temperature	Automatic meteorological station at Agia Paraskevi	Daily	(01/01/1955– 01/01/2020)	Data gaps filled with Mytilini airport data corrected by the average monthly difference between this station and that of Agia Paraskevi
Stream Discharge	Telemetric Radar Level Sensor (RLS) gauging station Prini-bridge	Daily	(08/01/2014– 10/31/2019)	Observed gauge data; Data gap: 11/01/2016–11/01/2017
Landuse	CORINE 2000 [34]	-	-	The map is corrected with the inclusion of five settlements and data gathered by field validation. 1:25,000
Soil	Municipality, Hellenic Survey of Geology and Mineral Exploration (HSGME), field sampling	-	-	Combined soil data from maps provided by the municipality, the HSGME and field sampling [35]
Topography	NASA Shuttle Radar Topography Mission (SRTM) Version 3.0	-	-	$1° \times 1°$ tiles at 1 arc second (about 30 m)

2.3. Regional Climate Model Data

The use of RCMs is necessary in regions with multiple topographic characteristics, and Greece is a Mediterranean country characterized by complex topography with steep orography from the mountainous regions to the coast, with elongated coastline and a number of small islands in the Aegean Sea and Ionian Sea. Hence, the use of dynamical downscaling to higher resolution is necessary to assess the regional and sub-regional climate of the complex topographically area of Greece [36,37].

Data series for several meteorological parameters (maximum and minimum temperature, precipitation, wind speed, and incoming shortwave radiation) of multiple RCMs, for Tsiknias basin over the time period 2021–2100, were used (Table 2). The projections are in a high resolution (0.11 deg) from various RCMs and emission scenarios (RCPs 4.5 and 8.5) based on data from EURO-CORDEX initiative [38]. EURO-CORDEX is the European branch of the international CORDEX initiative, which is a program sponsored by the World Climate Research Program (WRCP) to organize an internationally coordinated framework and produce improved regional climate change projections for all land regions worldwide. In this study, CORDEX results serve as input for climate change impact assessment, within the timeline of the Fifth Assessment Report (AR5). The data extraction and analysis is in the framework of GEO-CRADLE [39], which aims to provide a user-friendly web application tool for climate change impact studies, for end users and policy makers on climate change adaptation strategies.

Table 2. CORDEX multi-model datasets abbreviation used in this study.

RCM Institution Name	RCM Institution Acronym	RCM Model Name	GCM Model Name	GCM Institution Acronym	Abbreviation Used in This Study
Climate Limited-area Modeling-Community	CLM com	CCLM4-8-17	CNRM-CM5	CNRM-CERFACS	CNRM-CM5_CCLM4-8-17
Centre national des recherches météorologiques	CNRM	ALADIN53	CNRM-CM5	CNRM-CERFACS	CNRM-CM5_ALADIN53
Koninklijk Nederlands Meteorologisch Instituut	KNMI	RACMO22E	EC-EARTH	ICHEC	EC-EARTH_RACMO22E
Institut Pierre-Simon-Laplace	IPSL-INERIS	WRF331F	CM5A-MR	IPSL-IPSL	CM5A-MR_WRF331F
Sveriges Meteorologiska och Hydrolo-giskaInstitut	SMHI	RCA4	CM5A-MR	IPSL-IPSL	CM5A-MR_RCA4
Climate Limited-area Modeling–Community	CLMcom	CCLM4-8-17	HadGM2-ES	MOHC	HadGM2-ES_CCLM4-8-17
Sveriges Meteorologiska och Hydrolo-giskaInstitut	SMHI	RCA4	HadGM2-ES	MOHC	HadGM2-ES_RCA4
Climate Limited-area Modeling–Community	CLMcom	CCLM4-8-17	MPI-ESM-LR	MPI-M	MPI-ESM-LR I_CCLM4-8-17
Max Planck Institute Magdeburg	MPI-CSC	REMO2009	MPI-ESM-LR	MPI-M	MPI-ESM-LR_REMO2009

In this work, the focus is on the changes between the future and control period, to assess future sensitivity of the hydrological regime to the climate change. Thus, no bias correction was conducted on the CORDEX RCMs meteorological datasets, since it would not have a quantitative added value; on the contrary, it may induce other uncertainties and mask the climate change effects from linear corrections in past and future, for a system that is not linear (hydrological regime). The applicability of bias correction approaches especially for extreme hydrological indicators, such as high flows, was found to have high impacts on their values and generally is still questionable [40].

3. Methodology

The methodology applied in this study is described in Figure 3 and includes the following steps: (1) setup of the SWAT model and calibration/validation using SWAT-Cup algorithm SUFI2 (see Sections 3.1.2 and 3.1.3); (2) simulation of historical and future flow using the calibrated model forced by the CORDEX datasets (see Section 3.1.4); (3) classification of the Tsiknias River regime using metrics that measure the relative permanence of temporary flow phases within TREHS model; and (4) assessment of the degree of alteration due to climate change of the temporary regime of Tsiknias River (see Section 3.2.2).

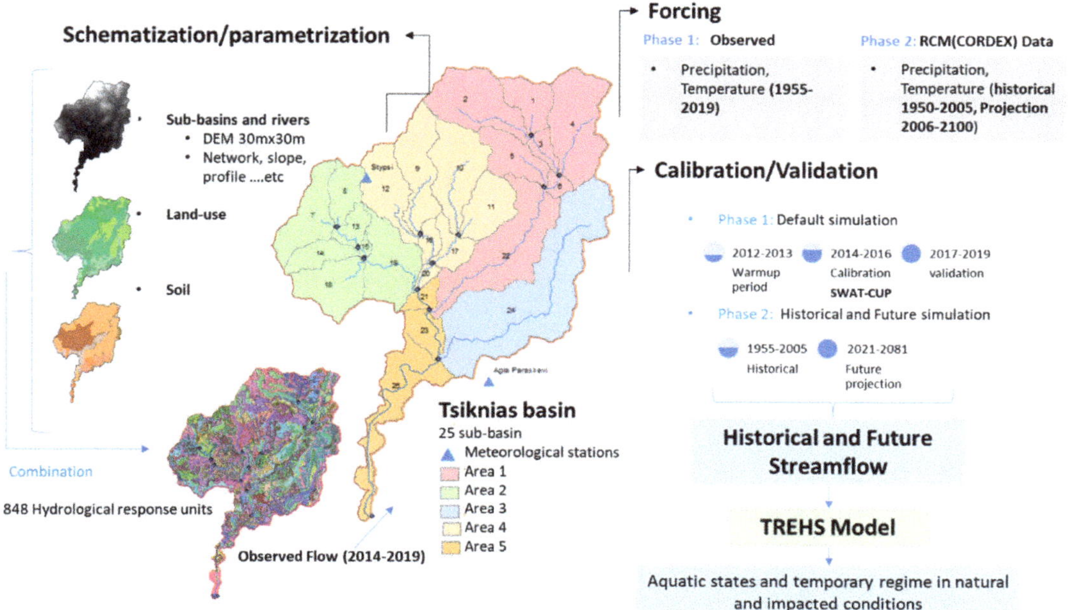

Figure 3. Flowchart of the overall methodology followed.

3.1. Hydrological Modeling

3.1.1. Soil and Water Assessment Tool (SWAT)

The SWAT model was chosen because of the plethora of parameters available for a better simulation of processes specific to each basin, its flexibility during the calibration stage, and the option of running climate change scenarios. SWAT [17–19,41], is a continuous, semi-distributed, physically-based hydrologic model, developed by the U.S. Department of Agriculture (USDA), to predict the impact of land management practices on water, sediments, and agricultural chemical yields in large complex basins with varying soils, land use, and management conditions over long periods of time [18]. It divides a basin into sub-basins connected by a stream network and further, divides each sub-basin into Hydrologic Response Units (HRUs) consisting of unique combinations of slope, land use, and soils. Runoff is predicted separately for each HRU and routed to obtain the total runoff for the basin. This increases the accuracy and gives a much better physical description of the water balance. SWAT model operates on a daily or sub-daily time-step for each hydrologic unit based on water balance equation and simulates the hydrology into land and routing phases. During the land phase, the amount of water, sediment, and other non-point loads are calculated from each HRU and summed up to the level of sub-basins. Each sub-basin controls and guides the loads towards the sub-basin outlet. The routing phase defines the flow of water, sediments, and other non-point sources of pollution through the channel network, from one sub-basin to another and to the outlet of the basin.

The hydrological cycle simulated by SWAT is based on the water balance equation:

$$SW_t = SW_0 + \sum_{i=1}^{t} \left(R_{day} - Q_{surf} - E_a - W_{seep} - Q_{lat} - Q_{gw} \right) \tag{1}$$

where SW_t is final soil water depth (mm), SW_0 is the initial soil water depth (mm), t is the time step (days), R_{day} is the daily precipitation (mm), Q_{surf} is a surface run-off (mm), E_a is the actual evapotranspiration (mm), Q_{gw} is the depth of water entering in vadose zone from soil profile (mm), and Q_{lat} is the depth of lateral flow (mm). Runoff is derived from the USDA Soil Conservation Service (SCS) runoff Curve Number (CN) method [42] as follows:

$$Q_{surf} = \frac{\left(R_{day} - I_a\right)^2}{\left(R_{day} - I_a + S\right)} \tag{2}$$

where R_{day} is rainfall depth for that day; I_a is the initial abstraction, which is a function of infiltration, interception and surface storage; and S is the retention parameter calculated from the curve number (CN), which is based on soil parameters and land use classes. CN is an important calibration parameter for surface runoff [19]. High CN corresponds to high overland flow often associated with developed soils, while low CN represents well-drained soils and results in low rates of surface runoff.

3.1.2. Model Setup

SWAT was setup for Tsiknias River basin through the ArcSWAT interface [43] using the land use, soil, and DEM datasets and meteorological data described in Table 1. The basin was divided into 25 sub-basins and 848 HRUs.

3.1.3. Model Calibration and Parametrization

The Tsiknias River basin was calibrated from 2014 to 2016 and validated from 2017 to 2019 on a daily time step with observed meteorological data and streamflow from Prini gauge station, with two years as a warmup period (2011–2013). The Sequential Uncertainty Fitting ver.2 (SUFI2) embedded within the SWAT Calibration and Uncertainty Program (SWAT-CUP) software [44–46] was employed for auto calibration, using the Nash–Sutcliffe Efficiency (NSE) as the objective function criterion. Nine parameters were selected (Table 3), based on dominant processes in Tsiknias basin reported by previous studies [26,47–50]. Previous streamflow measurements from July 2007 to July 2009 showed that the contribution of individual sub-basins, described in this work as Areas 1–5, respectively (Figure 4), were consistent with the size of the drainage area, the slope, and the existence of water springs [32].

The Percent Bias (PBias) [51] and NSE [52] were used to evaluate the agreement goodness of fit between observed and simulated data. In general, model simulation is regarded satisfactory if NSE > 0.50 and PBIAS = ±25% when simulated and observed streamflow are compared [53,54]. The parameter set that produced the best-fit stream discharge for the daily data was selected.

3.1.4. Future Streamflow Projections

After confirming SWAT model capability for hydrologic modeling during the previously selected time period (Section 3.1.3), CORDEX data (minimum and maximum air temperature, precipitation, wind speed, incoming shortwave radiation, and relative humidity) were applied to the calibrated SWAT model for the control period 1950–2005 to simulate historical monthly flow. Finally, CORDEX RCMs projected climate data under RCP 4.5 (intermediate scenario) and RCP 8.5 (worst-case scenario) [5] were used for simulating future monthly streamflow projections. The future time period of the simulation is 50 years (i.e., 2021–2071), having the same length as the control time period (i.e., 1955–2005).

Table 3. Parameters used in the calibration process.

Parameters [1]	Definition	Physically Meaningful Range (min max)		Calibration Range
r__CN2.mgt	Initial SCS runoff curve number for moisture condition	35	98	−50% to 20%
v__ALPHA_BF.gw	Base flow travel time (days)	0	1	0.2−1
v__GWQMN.gw	Threshold depth of water in the shallow aquifer required for return flow to occur (mm)	0	5000	−1000–2000
r__SOL_AWC.sol	Available water capacity of the soil layer (mm/mm)	0	1	−20% to 20%
v__ESCO.hru	Soil evaporation compensation factor	0	1	0.6–1
v__REVAPMN.gw	Threshold depth of water in the shallow aquifer for revap to occur (mm)	0	500	0–500
v__GW_REVAP.gw	Ground water revap coefficient	0.02	0.2	0.02–0.2
v__LAT_TTIME.hru	Lateral flow travel time (days)	0	180	0–150
r__SLSUBBSN.hru	Average slope length (m)	10	150	−25%–25%

[1] v__ means the default parameter is replaced by a given value within calibration range; r__ means the existing parameter value is multiplied by a given value; mgt, crop cover management process; gw, groundwater process; sol, soil water dynamics process; bsn, basin scale; rte, water routing; hru, water dynamics at HRU level.

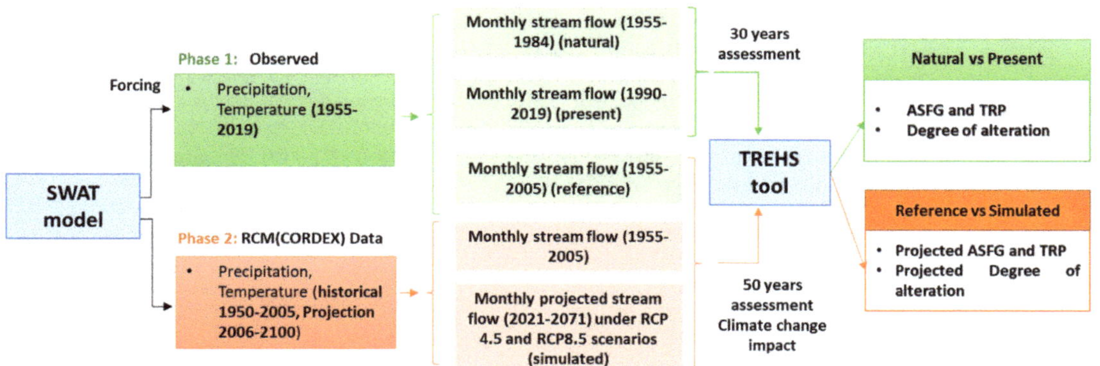

Figure 4. Flowchart of temporary regime assessment.

3.2. Aquatic States (AS), Flow Regime (FR) Status, and Alteration Flow Regime (AFR) Assessment

In this section, the aquatic states are defined, TREHS tool is described (Section 3.2.1), and the methodology followed to identify the different distribution of AS, FR, and AFR for Tsiknias River due to climate change effect is analyzed (Section 3.2.2).

3.2.1. Temporary Rivers Ecological and Hydrological Status (TREHS)

TREHS was developed as a tool for applying the methods formulated during the MIRAGE Project [14], aiming to help managers capture the temporary stream dynamic and discover the convenient methods to define hydrological and ecological status [16,55,56]. By means of updated methods and visualization options, the tool facilitates the assessment of the hydrology of temporary rivers. Diverse types of hydrological-related data sources can be used to define the specific regime of temporary rivers, the aquatic states and the degree of their alteration [16,57]. for instance, monthly flow data obtained from monitoring stations, or in the case of absence of observed data, which is the case in most of intermittent rivers, rainfall–runoff model simulations, terrestrial photography, and/or interviews of locals.

TREHS defines six different AS, which correspond to the transient sets of aquatic mesohabitats occurring on a given river reach at a particular moment, depending on the hydrological conditions [14,16,56,58], allowing a better evaluation of biological assemblages [14]. From wet to dry ASs are classified as follows: flooding conditions (Hyperrheic); full prevalence of all the possible mesohabitats (Eurheic); sequence of pools connected by flowing water threads (Oligorheic); occurrence of isolated pools (Arheic); disappearance of surface water, with the wet alluvium still allowing underground aquatic life (Hyporheic); and the desiccation of the riverbed and alluvium, involving the disappearance of any active aquatic habitat (Edaphic).

The identification of the temporal patterns of occurrence of these ASs is determined based on the statistics of the occurrence of these diverse aquatic states [57,58]. The flowing water phase (Hyperrheic, Eurheic, and Oligorheic) is separated from the zero flow phase (Arheic, Hyporheic, and Edaphic) using flow threshold values, which can be identified by field observations of the ASs for more accuracy or automatically in TREHS tool by the shape of flow duration curve.

This flow duration analysis, which also characterizes the ability of the basin to provide flows of various magnitudes [59] in a given period, is identified by ranking the flow data from highest to lowest and assigning an exceedance probability, P, to each value according to the following formula:

$$P = \frac{m}{n+1} \tag{3}$$

where P is the probability of exceedance, m is the rank of the data value (m = 1 being the largest), and n is the total number of data points.

The six metrics reflecting patterns of flow, isolated pools, and dry beds, for each data source are defined as follows: flow permanence (Mf), dry channel permanence (Md), isolated pool permanence (Mp), seasonal six-month predictability of the period without flow (Sd6), equinox-solstice seasonality (ESs), and summer-winter seasonality (SWs) [56,58]. The most important metrics are Mf and Sd6. They are used for the classification of flow into positive and zero flows, which have the most impact on the ecosystems of the stream [16,56,58]. Mf represents the long-term mean annual relative number of months with flow, which ranges between 0 (always dry) and 1 (always flowing). Sd6 signifies the seasonality of the dry conditions and hence the predictability of habitat availability, and it is described by the following equation [16].

$$Sd6 = 1 - \sum_{1}^{6} Fd_i \left/ \sum_{1}^{6} Fd_j \right. \tag{4}$$

where Fdi is the multi-annual frequency of zero-flow month i for the contiguous six wetter months of the year and Fdj is the multi-annual frequency of zero-flow month j for the remaining six drier months.

3.2.2. ASs, FR, and AFR Assessment Workflow

In summary the methodology followed to evaluate and determine the AS and the temporary FR in both natural and impacted (climate change) conditions of the Tsiknias temporary regime, is described Figure 4.

In the first part (Figure 4), natural AS (1955–1984) and present AS (1990–2019) were evaluated, using simulated monthly flow at one station scale, forced by observed meteorological datasets, and then were compared to each other to determine the degree of alteration. These were visualized by three plots provided by TREHS tool: (a) the Aquatic States Frequency Graph (ASFG) that shows the relative importance of the diverse states throughout the year and the degree of seasonality of the regime; (b) the Temporary Regime Plot (TRP) which presents the flow permanence against seasonal predictability in order to compare the occurrence of flow for different rivers (Permanent (P), Intermittent–Permanent (I-P), Intermittent–Dry (I-D), and Ephemeral (E)); and (c) the Flow-Pools-Dry plot (FPD) using the flow, pool, and dry permanence indicators [16].

Finally, using the projected monthly stream flows under RCP 4.5 and RCP 8.5 scenarios, we estimated the projected impacted aquatic states and their alteration. Similarly, to Part 1, results are presented in ASFG, TRP, and FDP plots.

4. Results

4.1. Hydrologic Modeling

In this section, the calibration and validation results of the SWAT hydrological model are discussed. Table 4 shows the obtained calibrated fitted value for each parameter for Tsiknias basin, while Figures 5 and 6 illustrate the resulting hydrographs of calibration and validation process at a daily time step.

All calibrated parameters are within the expected range for Tsiknias basin, with the most sensitive parameters being CN2, SLSUBBSN, and ESCO considering their p values. The model sensitivity to SLSBBSN confirms a previous study indicating that the size of the drainage area and the slope have great impact on the outflow within this basin (the highest elevation is 943 m and the lowest is −1 m) [32]. The small final fitted range of CN2 indicates low run-off potential of the basin due to land-use coverage, mostly agricultural (olive groves), significant natural cover such as coniferous forests in the northern part of the basin, grassland, and brushland habitats [48].

Table 4. Parameters used in calibration process.

Parameter	Best Simulation Fitted Value	p Value	Final Range (min, max)
r__CN2.mgt	−0.08	0.00	−0.15, −0.008
r__SLSUBBSN.hru	−0.435	0.01	−0.8, −0.08
v__GWQMN.gw	0.26	0.04	−0.17, 0.7
v__ESCO.bsn	0.783	0.05	0.72, 0.83
v__ALPHA_BF.gw	0.67	0.32	0.5, 0.8
r__SOL_AWC.sol	−0.005	0.39	−0.05, 0.05
v__LAT_TTIME.hru	16.65	0.46	0, 57.6
v__REVAPMN.gw	113.75	0.5	16.9, 210.6
v__GW_REVAP.gw	0.122	0.72	0.1, 0.15

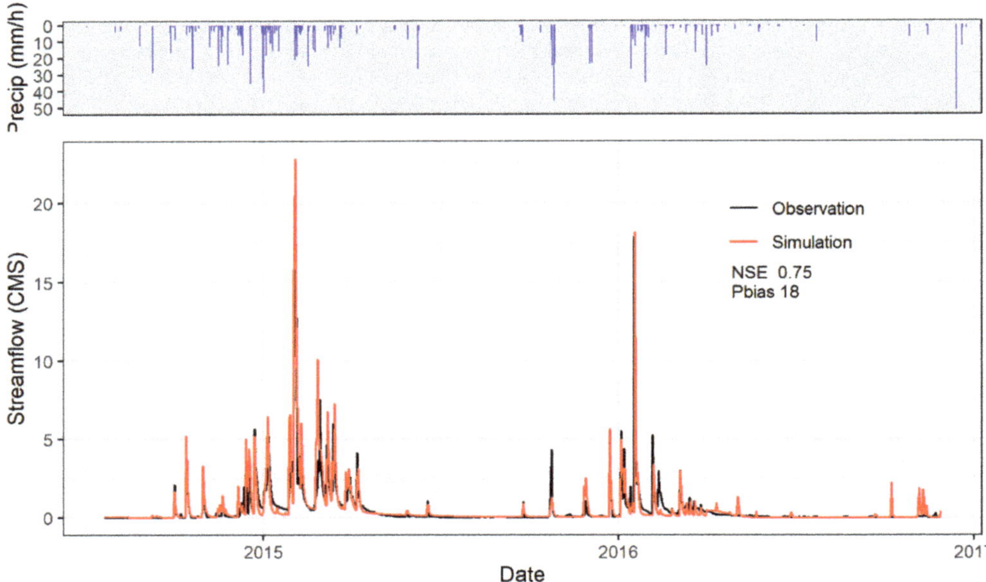

Figure 5. Measured and simulated daily discharge at Tsiknias River gauging station during calibration (2014–2016).

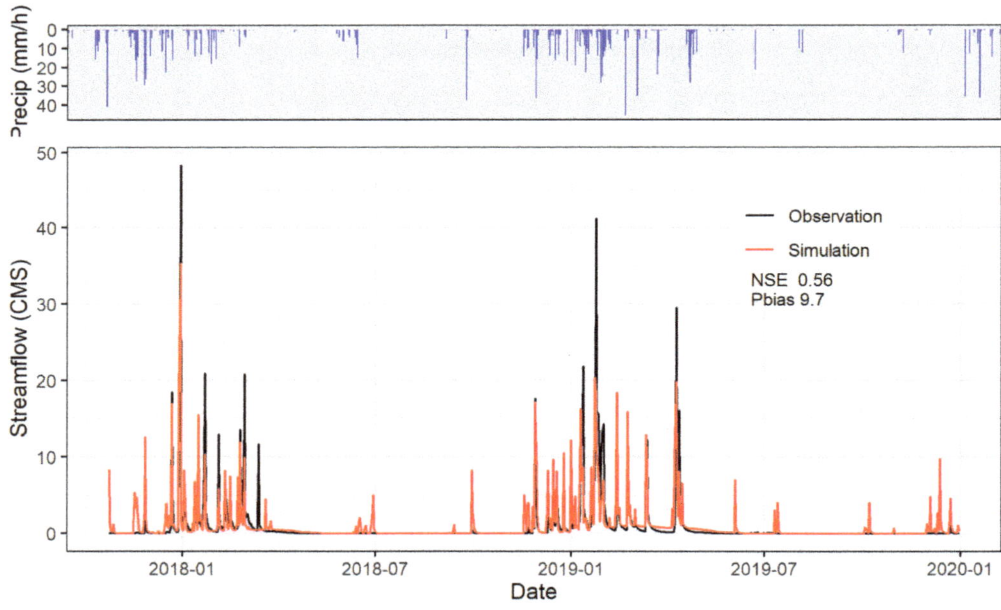

Figure 6. Measured and simulated daily discharge at Tsiknias River gauging station during validation (2017–2019).

Statistical evaluations of both calibration and validation, as shown in Figures 5 and 6, are within the acceptable levels reported in the literature (NSE values > 0.5, Pbias < 25%) [54]. The results show a positive correlation between the observed and simulated river flows and the water mass balance of Tsiknias River at annual scale. The simulated flow was slightly underestimated during the autumn; it sharply increases during the winter and levels off after the end of the wet period. SWAT, however, does not entirely capture both peaks in February 2015; it slightly overestimates it (Figures 5 and 6). The validation flow records represent a time period during which the gauging and meteorological station are not adequately preserved, and sediment and debris have altered the examined river cross section.

Since the TREHS model requires monthly data to assess the temporary regime aquatic state and alteration, keeping the same calibrated parameters, SWAT model was run on a monthly timestep. Figure 7 shows the resulting hydrograph during the 2014–2019 period. The contribution of the different tributaries of the basin [32] was evaluated, compared to the total basin discharge, and confirmed. The main tributary referred to as Area 1 (Figure 2) is the main water contributor during the wet period (Figure 7, Subbasin 22 outflow) due to the high elevation difference, its great drainage area (30% of the basin), and the existence of spring sources. The secondary tributary (Area 4, Subbasin 20 outflow, 20% of total surface of basin) and the ephemeral tributaries (Areas 3 (Subbasin 24) and 2 (Subbasin 19), covering 15% and 17%, respectively) equally contribute less water to the main stream (Area 5, Subbasins 23 and 25 outflow). The ephemeral tributaries are showing a steeper reduction during spring in the end of the wet season, attributed to their ephemeral nature. The discharge at the Prini station of Tsiknias River corresponds to the sum of the outflows from the contributing streams. Correspondingly, the low slope at the lowland area near the outlet allows for higher infiltration with regards to the upstream parts of the basin and experiences higher groundwater recharge rates, resulting in zero flow during the dry period.

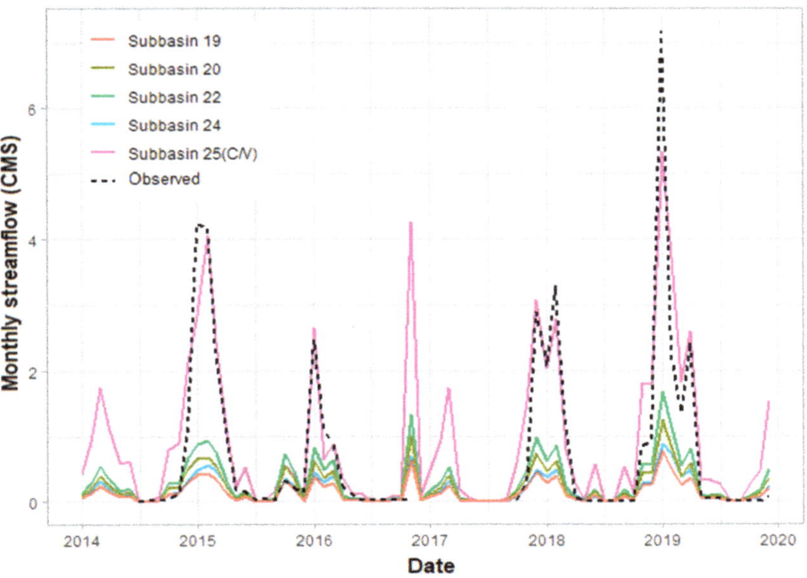

Figure 7. Measured and simulated monthly discharge at Tsiknias River contributing tributaries during 2014–2019.

4.2. Future Climate

Based on high resolution regional climate data (0.11°) acquired from CORDEX, the multi-model simulated time series of changes in the annual precipitation and temperature relative to the period 1961–1990 for the scenarios RCP 4.5 and RCP 8.5 as well as the historical period are shown in Figures 8 and 9. The lines indicate ensemble yearly means from the yearly ensemble mean data. All precipitation scenarios for the future indicate a slight but not robust decrease of precipitation, while temperature projections show warming trends for the different RCPs (highest for RCP 8.5 and lowest for RCP 4.5) which drift apart significantly during the second half of the 21 century. During the same period, the ensemble standard deviation (shaded area) is increasing, indicating a rise in the uncertainty of the simulated ensemble temperature and precipitation.

The projected climate variable of rainfall for the future period (2021–2071) under both RCPs was analyzed against the baseline situation (1955–2005). The precipitation mean monthly cycles are presented in Figure 10 for the historical period, RCP 4.5, RCP 8.5, and observed. Both RCP 4.5 and RCP 8.5 scenarios project a decrease in precipitation mostly during the wet season.

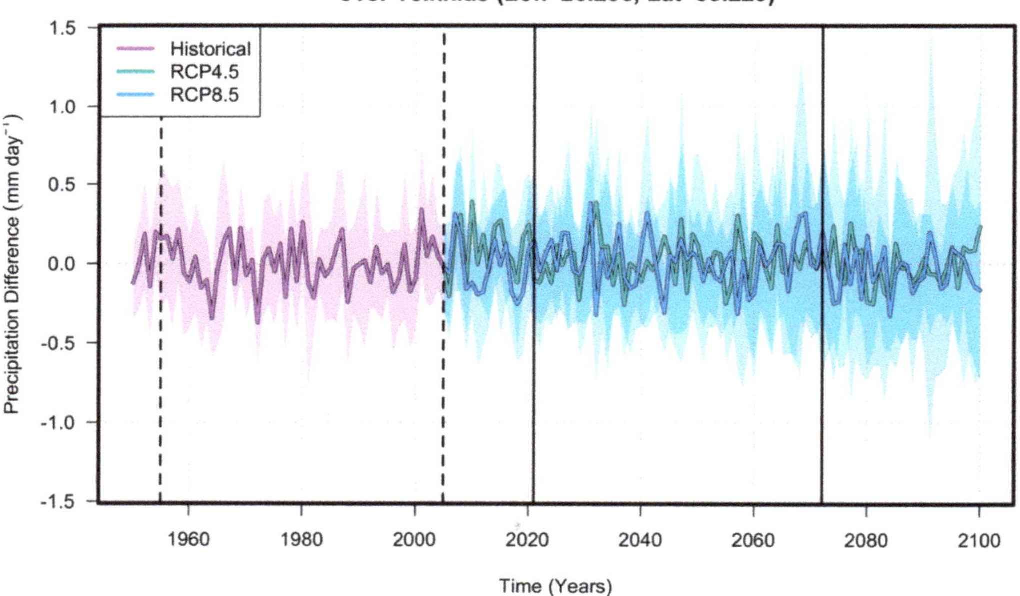

Figure 8. Multi-model simulated time series at Tsiknias (Lesvos) from 1950 to 2100 indicating the changes in annual precipitation relative to the period 1961–1990 for the historical, RCP 4.5, and RCP 8.5 experiments. Lines and shades represent the ensemble mean and standard deviation of all the available CORDEX models, respectively. Vertical dashed and continuous lines delimit the periods 1955–2005 and 2021–2071, respectively.

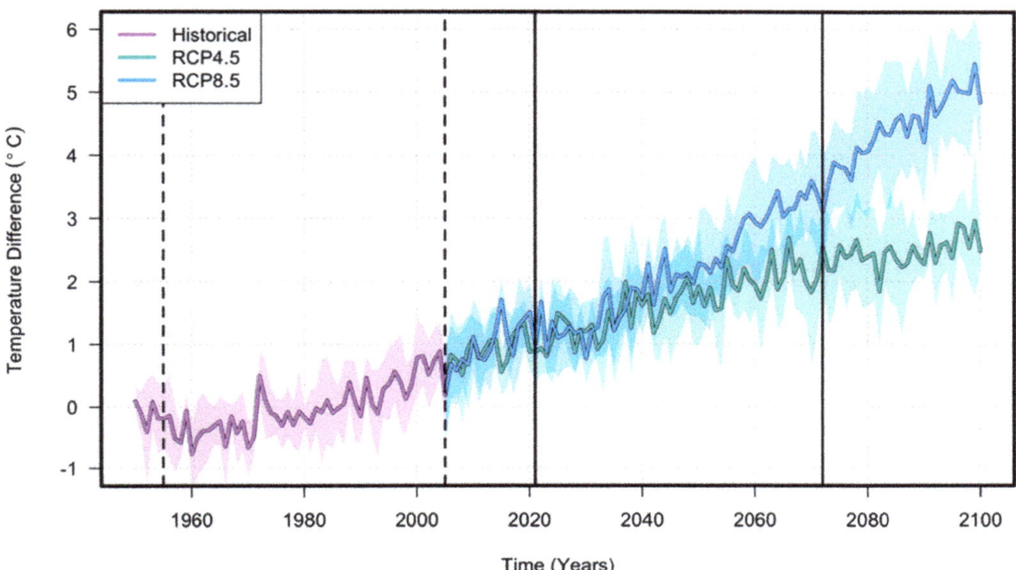

Figure 9. Multi-model simulated time series at Tsiknias (Lesvos) from 1950 to 2100 indicating the changes in mean annual near surface temperature relative to the period 1961–1990 for the historical, RCP 4.5, and RCP 8.5 experiments. Lines and shades represent the ensemble mean and standard deviation of all the available CORDEX models, respectively. Vertical dashed and continuous lines delimit the periods 1955–2005 and 2021–2071, respectively.

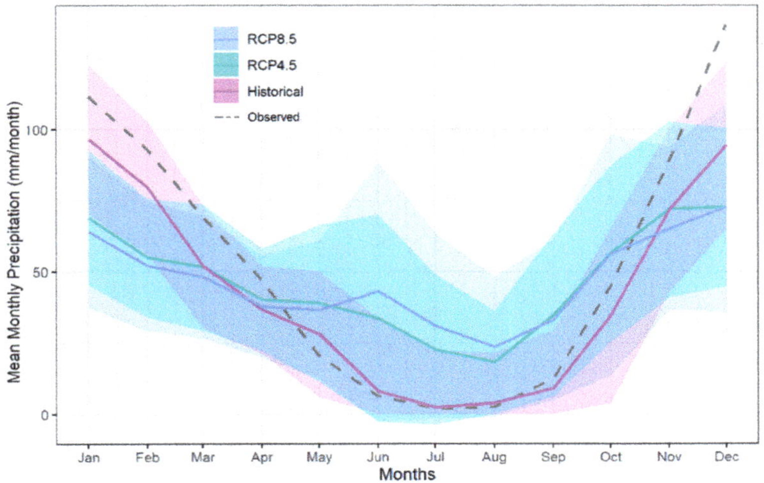

Figure 10. Multi-model simulated time series at Tsiknias (Lesvos) from 1950 to 2100 indicating the changes in mean monthly precipitation relative to the period 1961–1990 for the historical, RCP 4.5, and RCP 8.5 experiments, and observed 1955–2005. Lines and shades represent the ensemble mean and standard deviation of all the available CORDEX models, respectively.

4.3. Future Streamflow Projections (Monthly and Annual)

Figure 11 presents boxplots of the streamflow distributions of the control period (1955–2005) and future streamflow projections (from 2021 to 2071 under RCP 4.5 and 8.5 scenarios) forced by CORDEX RCMs datasets listed in Table 2 compared with the reference period (1955–2005) simulated by SWAT model. How well different RCMs capture characteristics of the reference flow distribution and the change future scenarios is indicated. Primary evaluation of the simulated outflow for the historical climate (1955–2005) against simulated streamflow shows that the annual and monthly observations are captured well by the ICHEC_KNMI model, unlike other RCMs. Analysis of the future scenarios (2021–2071) projections of the RCM/GCM ensembles for all models shows an increase in extreme events occurrence. Ultimately, almost all RCMs show a decrease in flow in the future and higher occurrence of extreme flood events highlighted by the presence of outliers (Figure 11).

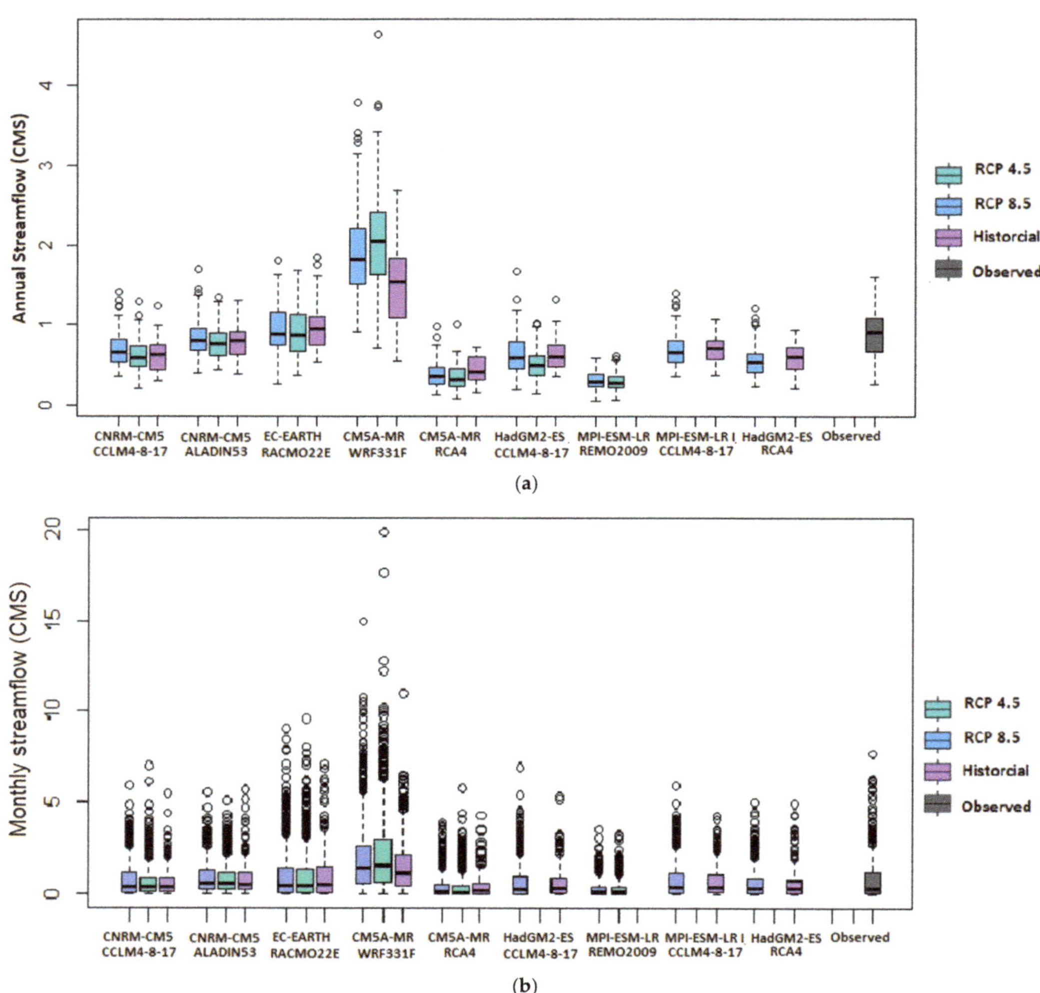

Figure 11. Box-and-whiskers plots for monthly (**b**) and annual (**a**) streamflow of Tsiknias River for each RCM/GCM combination. The central line is the median, while whiskers represent 25th and 75th percentiles.

In Figure 12, the actual flow duration curve of Tsiknias streamflow is compared to flows generated from RCM simulated flow. Most models generally replicate the flow pattern, yet exhibit major differences, particularly in low flows. As for the future, a higher frequency of low flows is projected, as well as an increase of the frequency of mid-range and high flows (Figure 12).

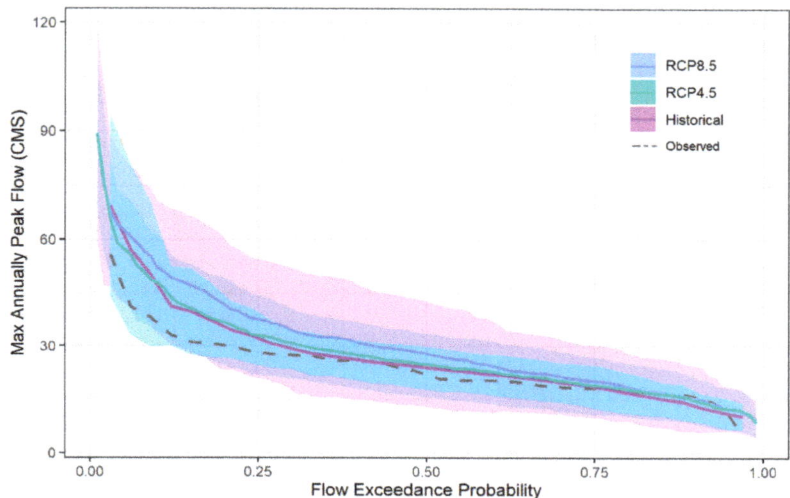

Figure 12. Flow duration curve of reference data from 1955 to 2005 against historical scenarios and projected future flow duration curve under RCP 4.5 and RCP 8.5 scenarios.

4.4. Hydrological Status

Tsiknias River is characterized by a temporary hydrological regime subject to seasonal abstraction in order to cover irrigation needs; it is, therefore, prone to more alteration [26]. Assessment of the current and projected future flow regimes, however, is important to understand the existing ecosystems functioning and river processes sustainability. In this section, the classification of Tsiknias flow regime and the distribution of the aquatic states is evaluated firstly, under natural (1955–1984) and present (1990–2019) conditions. Then, using CORDEX RMCs simulated monthly flows, the projected conditions (2021–2071) are gauged.

4.4.1. ASFG (Natural, Actual, and Projected)

The ASFG describes the relative frequency (%) of the aquatic states through the year, which is derived from the monthly frequencies of the simulated flows. It is generally applied for the selection of the most appropriate sampling methods (e.g., biomonitoring) according to the temporal variability and seasonal predictability of the transient sets of aquatic mesohabitats occurring in a stream [56], but also for the visualization of the flow regime of the streams. The threshold values between flow phases were fixed automatically in TREHS based on the flow duration curve. The ASFG was first generated for Tsiknias River during the natural state (1955–1984) and actual state (1990–2019) (Figures S1–S3), natural state represents the period with the minimum human intervention to the flow regime and the absence of any irrigation projects.

Figure 13 shows ASFG obtained from reference discharge (1955–2005) simulated by the calibrated SWAT model. Under natural state, Tsiknias River exhibits clear seasonality (dry summer and relatively wet winter). The permanence of the flow is considerable (Eurheic 83.7%, indicating full prevalence of all the possible mesohabitats), although it dries more than 10% of the time.

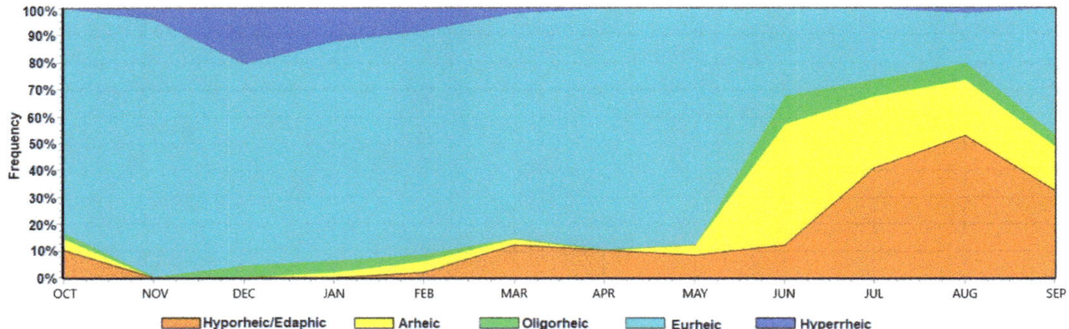

Figure 13. ASFG of the natural state of Tsiknias River derived from flow simulations (1955–2005).

After describing the actual flow regime of Tsiknias River, its assessment under different scenarios, RCP 4.5 and RCP 8.5, can be useful for decision makers for taking appropriate measures to mitigate the alteration of flow regimes due to the impact of climate change. The same methodology was used during the assessment of the aquatic state frequency. Figure 14 shows a set of ASFGs obtained from diverse RCM simulations under both scenarios (RCP 4.5 and RCP 8.5) to determine the possible scenarios of alteration of the aquatic states due to climate change. According to the ASGF graphs (Figure 14), most RCMs indicate that Tsiknias River will be affected by increasing dry state and low flows periods under both RCP 4.5 and RCP 8.5 scenarios. CM5A-MR_RCA4 and MPI-ESM-LR_REMO2009 models on the other hand show a different pattern and loss of seasonality. HadGM2-ES_CCLM4-8-17 and HadGM2-ES_RCA4 models show a different pattern even in the historical period.

4.4.2. Projected Temporary Regime Alteration

The metrics of flow permanence, Mf, and seasonal predictability of dry periods, Sd6, were first evaluated for Tsiknias River in unimpacted (natural) and impacted conditions (climate change) using the SWAT simulated monthly outflows. The results are shown in TRP graphs (see the Supplementary Materials). They provide a classification of the river types, where the intermittence of the river increases from the upper right corner to the lower left. Gallart et al. (2012) [56] defined the limits between regime types in the Mediterranean region by analyzing flow time series from different streams within this area, which are Permanent (P), Intermittent–Permanent (I-P), Intermittent–Dry (I-D), and Ephemeral (E). They are differentiated by the grey lines on the graph. Furthermore, climate change impact on Tsiknias temporary regime was assessed by analyzing the shift between corresponding points of RCMs under both scenarios and natural conditions point. Analysis indicates the projected possible alteration degree of Tsiknias flow regime, due to the changes occurring in flow permanence (Mf) and dry season predictability (Sd6) (Figure 15). The results in the TRP indicate that the actual state of Tsiknias stream is Intermittent–Permanent (I-P) and the stream keeps this aquatic state according to many climate change scenarios. Both RCP 4.5 and RCP 8.5 scenarios predict a shift in the aquatic state to I-P. The distance between the corresponding points in unimpacted and impacted conditions is an indicator of the hydrological regime alterations capturing a shift in flow permanence. The FPD graphs are divided into nine aquatic phase regime classes based on three metrics corresponding to the proposed three aquatic phases: flow permanence, isolated pools permanence, and dry river permanence [16]. According to the reference period, Tsiknias River exhibits an alternate fluent regime ($0.40 < Mf \leq 0.90$, $0.00 \leq Mp < 0.50$, and $0.10 \leq Md < 0.60$), meaning that it rotates between three aquatic phases. In impacted conditions, however, most models under both RCPs scenarios generally indicate an alternate–fluent regime with a decrease in dry and pool permanence (Figure 16). CM5A-MR_WRF3331F and CNRM-CM5_ALADIN53 models exhibit a shift from a quasi-perennial regime in reference conditions to an alternate–fluent aquatic regime. HadGM2-ES_CCLM4-8-17 and HadGM2-ES_RCA4 models indicate a shift in the opposite direction under RCP 4.5 and RCP 8.5.

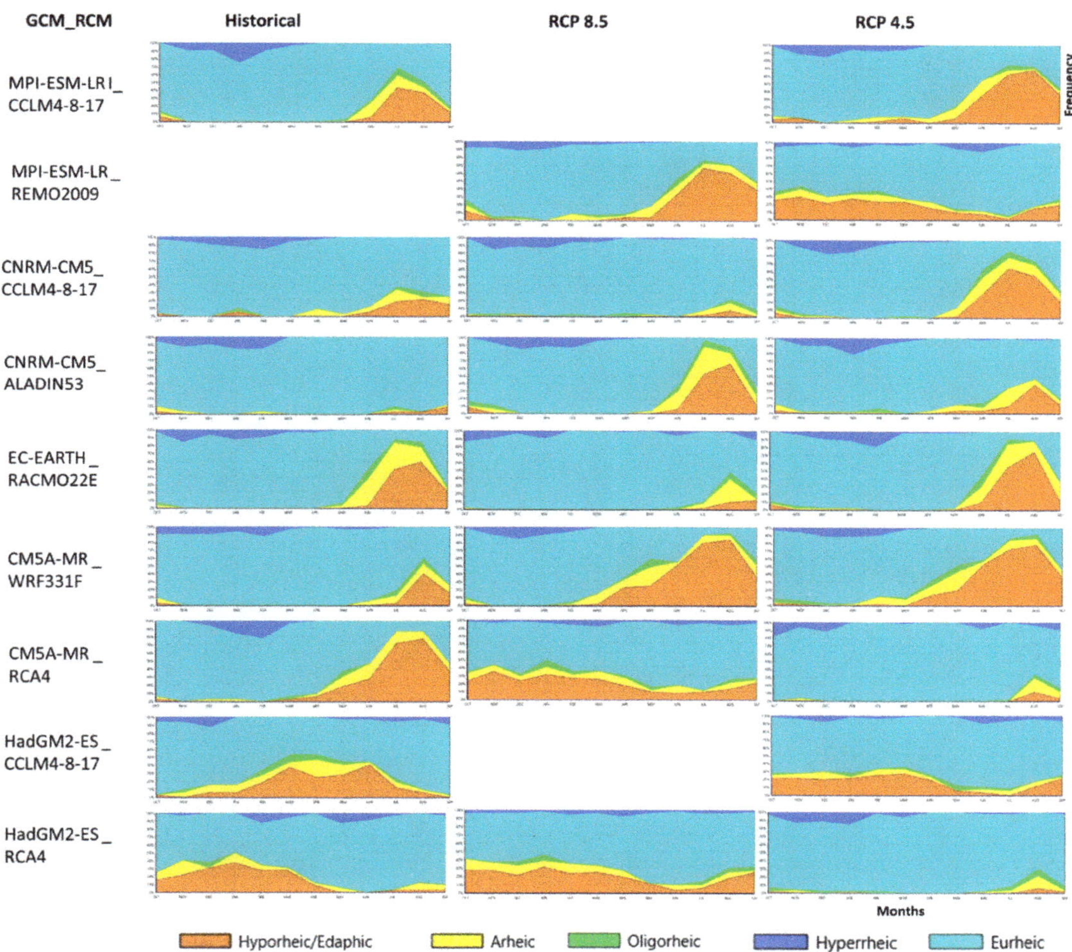

Figure 14. Aquatic states frequency graphs of historical scenarios: from 1955 to 2005 (**left**); and from 2021 to 2071 under RCP 8.5 scenario (in the **middle**) and RCP 4.5 (**right**).

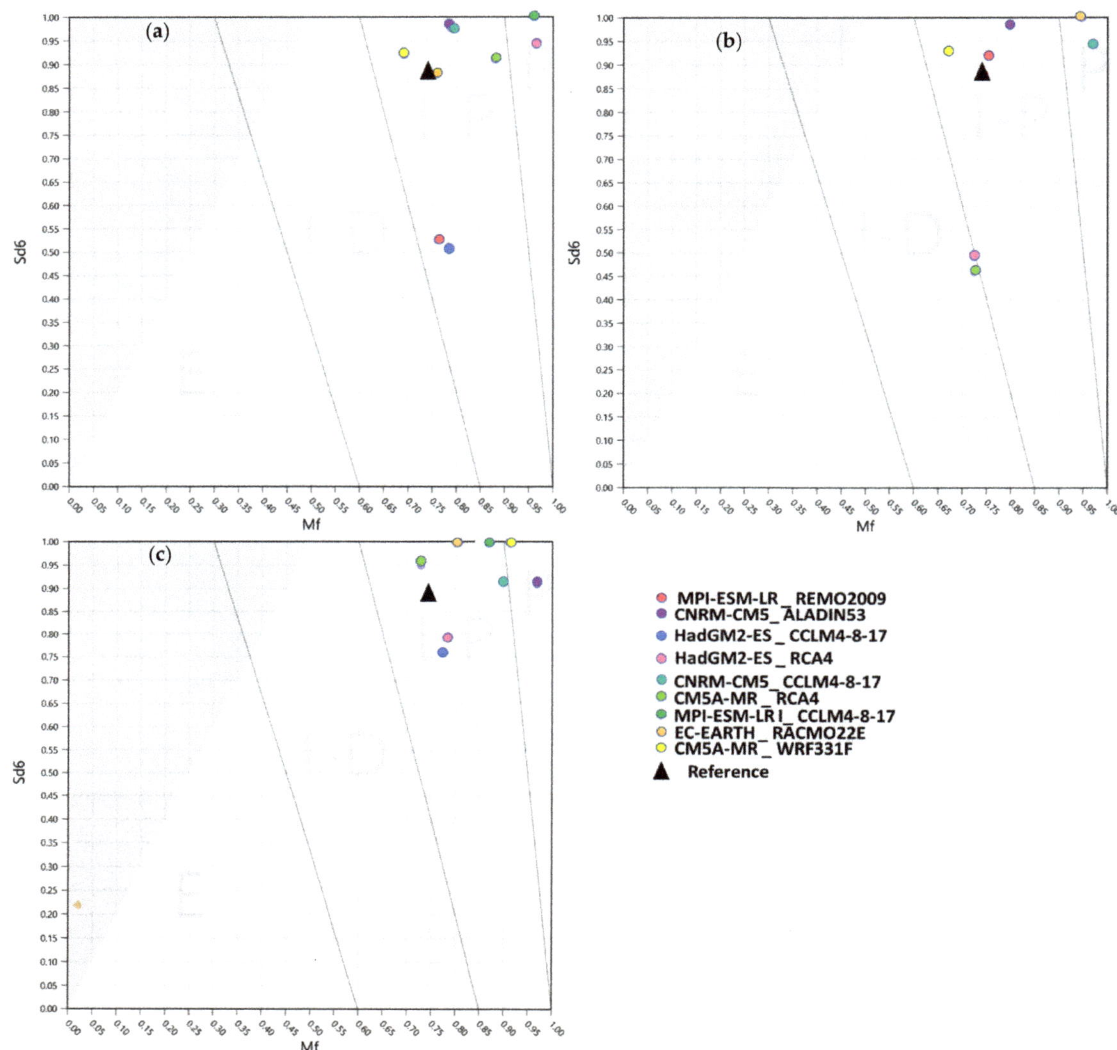

Figure 15. TRPs for the diverse sources: (**a**) RCP 4.5; (**b**) RCP 8.5; and (**c**) historical.

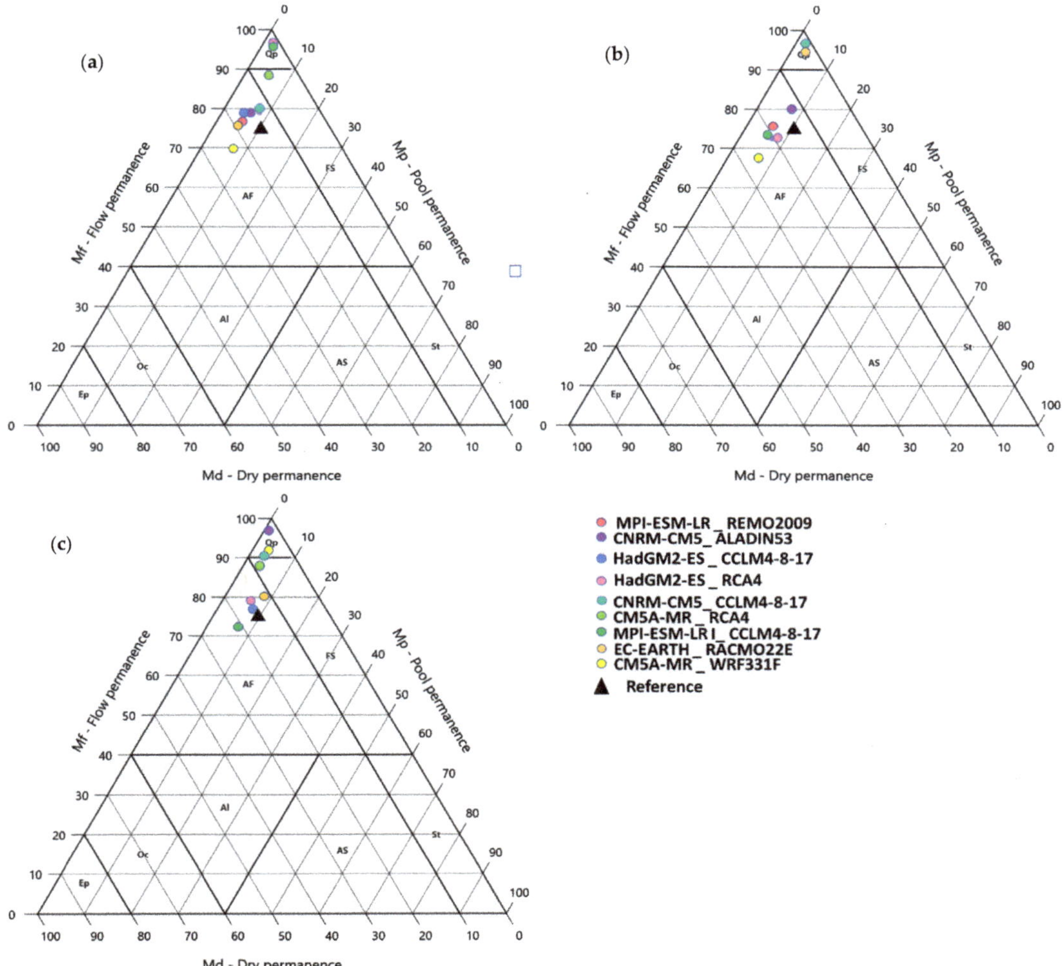

Figure 16. FDP plots for the diverse sources: (**a**) RCP 4.5; (**b**) RCP8.5; and (**c**) historical. Perennial (Pe), Quasi-perennial (Qp), Fluent-Stagnant (FS), Alternate-Fluent (AF), Stagnant (St), Alternate-Stagnant (AS), Alternate (Al), Occasional (Oc), Episodic (Ep).

5. Discussion

CORDEX RCM data were used to assess the climate change impact on the Tsiknias River hydrological regime. The results indicate that Tsiknias basin will suffer a combination of increased temperature and slightly reduced rainfall that will directly impact the flow regime; similar findings have been found in different parts of Mediterranean region [20,60–62]. The results of some RCMs simulations, however, do not capture the observed flow seasonality, which calls for further analysis, as it will provide further insight into projected climate model data and improve the accuracy of results, therefore improving the climate change impact adaptation strategies [60].

Analysis of the difference between the flow regime and distribution of its different aquatic states, based on the monthly average river discharge, the permanence of flow, and the six-month dry season predictability, between natural and present conditions, displays little to no alteration; nevertheless, in a neighboring watershed in Italy [62], the alteration was observed mostly downstream of the studied basin. Projected temperature increase and precipitation decrease, however, are expected to induce a change in the future flow regime, which exhibits an increase in summer and decrease in winter of flood events. These results are echoed in other studies within the Mediterranean region [63,64]; the predictions indicate high flow magnitudes will increase, the dry season will be extended, and extreme low flow conditions will be more aggravated. Furthermore, the major threat on the ecohydrological regime is expected in the last 30 years of the twenty-first century [63].

Using hydrological modeling for the reconstruction of the streamflow within an ungauged river basin and over the historical period is useful when observed data are missing [10]. In that scope. SWAT model was able to simulate streamflow in temporary river system, and therefore used to generate future flows to assess the projected hydrological regime of Tsiknias River. The variability of the aquatic states and thus ecosystems of temporary rivers make them more prone to be severely threatened by hydrological alteration [12]. Tsiknias River was identified as a temporary stream, which is the case of more than 50% of Mediterranean rivers [65]. Furthermore, it was concluded that high and low flows will become more extreme and dry periods will extend for longer periods as a result of climate change impact under both scenarios RCP 4.5 and RCP 8.5. These results resonate with different studies conducted in Europe [66] and the Mediterranean region in particular [6,12,67–69].

TREHS tool was tested and subsequently applied to assess the hydrological regime of temporary regime, however the ecological response to hydrological alterations was not evaluated, due to lack of data. Additionally, this model was previously tested on 119 stations from the Catalan River basin district, the Júcar River basin district, and the Ebro River basin district in Spain by the developers [16]. The results of these studies can be found in the TREHS database, allowing a comparison with other stations within the Mediterranean region. However, due to some limitations in the visualization of the results, this feature was not explored. Furthermore, using other methods offered by the TREHS tool such as photometry, observations, and interviews, is still to be explored. This tool is very useful and can be more advantageous due to rapid technological advancements in sensors and the massive penetration of smartphone technologies, which could facilitate the engagement of citizens in data collection in the context of hydrology, thus enhancing the long-term sustainability of monitoring networks [70].

6. Conclusions

In the present study, the assessment of flow and possible alteration of the hydrological and ecological status due to climate change impacts on the intermittent stream of Tsiknias stream was addressed, using hydrological modeling and ecological assessment. SWAT model with its plethora of available calibration parameters proved flexible enough during calibration and was able to reproduce the outflow of Tsiknias stream satisfactorily. Next, the calibrated model was run for historical (1955–2005) and projected (2021–2071) periods under RCP 4.5 and RCP 8.5 using CORDEX model simulations. The predicted increase in temperature and decrease in precipitation in the future resulted in a decrease of flow in the future and higher occurrence of extreme flood events captured with almost all RCM model simulations.

Using the flow records and RCMs simulated by SWAT flows, the present and projected four ASs of Tsiknias River were determined using the TREHS model. The ASFG of the actual flow is made exclusively with the use of the existing streamflow records without exploring the available options of aerial photography and interviews, which could have had a positive impact on the obtained results. The ASFGs showed high variability of the aquatic states in the future with a clear loss of the flow seasonality and an increase of the dry (Arheic and Hyporheic) states. The majority of the climate change scenarios predict an increase of the extreme flood events and expansion of the dry aquatic states of the IRES. The occurrence of extreme events is distributed not only to the winter months but in some cases to the whole year period. The ecosystem of this intermittent flow river is vulnerable to all these changes, with higher fragility in the flow-pool conditions, depicting future droughts, harming the local agriculture, ecological systems, and the socioeconomic life. Therefore, water retention measures could be considered such as the construction of reservoirs to store water and the adaptation of a set of nature-based solutions (i.e., small dams, weirs, and riparian restoration) to shield from future extreme flood phenomena.

IRES are complex hydrological systems. However, there are important obstacles for their proper study such as the hydrological or spatial information. The use of distributed models such as SWAT can facilitate the understanding of their response to the various human interventions and to the climate change impact. Finally, the application of the TREHS tool constitutes a great asset in defining the environmental flow requirements to prevent the degradation of these ecological pillars.

Supplementary Materials: The following are available online at https://www.mdpi.com/2306-533 8/8/1/43/s1. Figure S1: Aquatic State Frequency Graph (ASFG): (a) natural state (1955–1984); and (b) present (1990–2019); Figure S2: Temporary Regime Plot (TRP) of Tsiknias River under natural (red) and actual condi-tions (blue); Figure S3. Flow-Pools-Dry (FDP) plot of Tsiknias River under natural (red) and actual conditions (blue).

Author Contributions: Conceptualization, S.N. and O.T.; methodology, S.N., L.B. and O.T.; software calibration/validation, S.N.; Climate Change Scenarios, P.Z., T.T. and D.A.; and writing—review and editing, all authors. All authors have read and agreed to the published version of the manuscript.

Funding: N.S. acknowledges the ERASMUS + grant from the University of Aegean (Greece), which supported this work during the mobility period, and for their monitoring and control after the end of this mobility. O.T. received support for the instrumentation from the program the National Strategic Reference Framework (NSRF) 2014–2020, through the project "Observatory of Coastal Environment–AEGIS". Thanks to Prof K Kalabokidis (University of the Aegean) for the access to AEGIS platform dataset.

Institutional Review Board Statement: Not applicable.

Informed Consent Statement: Not applicable.

Data Availability Statement: Not applicable.

References

1. *The European Communites Directive 2000/60/EC of the European Parliament and of the Council of 23 October 2000 Establishing a Framework for Community Action in the Field of Water Policy*; Official Journal of the European Communities: Brussel, Belgium, 2000; Volume OJ L 327, pp. 1–73.
2. Datry, T.; Singer, G.; Sauquet, E.; Jorda-Capdevila, D.; Von Schiller, D.; Stubbington, R.; Magand, C.; Pařil, P.; Miliša, M.; Acuña, V.; et al. Science and Management of Intermittent Rivers and Ephemeral Streams (SMIRES). *Res. Ideas Outcomes* **2017**, *3*, e21774. [CrossRef]
3. Kaletová, T.; Loures, L.; Castanho, R.A.; Aydin, E.; da Gama, J.T.; Loures, A.; Truchy, A. Relevance of intermittent rivers and streams in agricultural landscape and their impact on provided ecosystem services—A mediterranean case study. *Int. J. Environ. Res. Public Health* **2019**, *16*, 2693. [CrossRef] [PubMed]
4. Datry, T.; Larned, S.T.; Tockner, K. Intermittent rivers: A challenge for freshwater ecology. *Bioscience* **2014**, *64*, 229–235. [CrossRef]
5. IPCC. *Climate Change 2013*; IPCC: Geneva, Switzerland, 2014; Volume 5, ISBN 9781107661820.
6. Loizidou, M.; Giannakopoulos, C.; Bindi, M.; Moustakas, K. Climate change impacts and adaptation options in the Mediterranean basin. *Reg. Environ. Chang.* **2016**, *16*, 1859–1861. [CrossRef]
7. Pachauri, R.K. *Climate Change 2014 Synthesis Report*; IPCC: Geneva, Switzerland, 2014; ISBN 9789291691432.
8. Giorgi, F.; Lionello, P.; Ictp, A.S.; Lecce, U. Climate Change Projections for the Mediterranean Region Submitted to Global and Planetary Change Special Issue on Mediterranean climate variability Abstract Key words: Climate change, Mediterranean climate, Precipitaton change, Temperature change. *Glob. Planet. Chang.* **2008**, *63*, 90–104. [CrossRef]
9. Alpert, P.; Krichak, S.O.; Sha, H.; Haim, D.; Osetinsky, I. Climatic trends to extremes employing regional modeling and statistical interpretation over the E. Mediterranean. *Glob. Planet. Chang.* **2008**, *63*, 163–170. [CrossRef]
10. Fonseca, A.R.; Santos, J.A. Predicting hydrologic flows under climate change: The Tâmega Basin as an analog for the Mediterranean region. *Sci. Total Environ.* **2019**, *668*, 1013–1024. [CrossRef]

11. Zittis, G.; Hadjinicolaou, P.; Klangidou, M.; Proestos, Y.; Lelieveld, J. A multi-model, multi-scenario, and multi-domain analysis of regional climate projections for the Mediterranean. *Reg. Environ. Chang.* **2019**, *19*, 2621–2635. [CrossRef]

12. Tzoraki, O.; Girolamo, A.D.; Gamvroudis, C.; Skoulikidis, N. Assessing the flow alteration of temporary streams under current conditions and changing climate by Soil and Water Assessment Tool model. *Int. J. River Basin Manag.* **2015**, 1–10. [CrossRef]

13. Cramer, W.; Guiot, J.; Fader, M.; Garrabou, J.; Gattuso, J.P.; Iglesias, A.; Lange, M.A.; Lionello, P.; Llasat, M.C.; Paz, S.; et al. Climate change and interconnected risks to sustainable development in the Mediterranean. *Nat. Clim. Chang.* **2018**, *8*, 972–980. [CrossRef]

14. Prat, N.; Gallart, F.; Von Schiller, D.; Polesello, S.; García-Roger, E.M.; Latron, J.; Rieradevall, M.; Llorens, P.; Barberá, G.G.; Brito, D.; et al. The Mirage Toolbox: An Integrated Assessmen TOOl for Temporary Streams. *River Res. Appl.* **2014**, *30*, 1318–1334. [CrossRef]

15. Gauthier, M.; Launay, B.; Le Goff, G.; Pella, H.; Douady, C.J.; Datry, T. Fragmentation promotes the role of dispersal in determining 10 intermittent headwater stream metacommunities. *Freshw. Biol.* **2020**, *65*, 2169–2185. [CrossRef]

16. Gallart, F.; Cid, N.; Latron, J.; Llorens, P.; Bonada, N.; Jeuffroy, J.; Jiménez-Argudo, S.-M.; Vega, R.-M.; Solà, C.; Soria, M.; et al. TREHS: An open-access software tool for investigating and evaluating temporary river regimes as a first step for their ecological status assessment. *Sci. Total Environ.* **2017**, *607–608*, 519–540. [CrossRef] [PubMed]

17. Santhi, C.; Srinivasan, R.; Arnorld, J.; Williams, J. A modeling approach to evaluate the impacts of water quality management plans implemented in a watershed in Texas. *Environ. Model. Softw.* **2006**, *21*, 1141e1157. [CrossRef]

18. Neitsch, S.; Arnold, J.G.; Kiniry, J.; Williams, J. *Soil and Water Assessment Tool Theoretical Documentation Version 2009*; Texas Water Resources Institute: College Station, TX, USA, 2009.

19. Arnold, J.G.; Moriasi, D.N.; Gassman, P.W.; Abbaspour, K.C.; White, M.J.; Srinivasan, R.; Santhi, C.; Harmel, R.D.; van Griensven, A.; Liew, M.W.; et al. SWAT: Model Use, Calibration, and Validation. *Trans. ASABE* **2012**, *55*, 1317–1335. [CrossRef]

20. Abouabdillah, A.; Oueslati, O.; Girolamo, A.M.D.; Porto, A.L. Modeling the impact of climate change in a mediterranean catchement (merguellil, Tunisia). *Fresenius Environ. Bull.* **2010**, *19*, 2334–2347. [CrossRef]

21. Papatheodouliu, A.; Tzoraki, O.; Panagos, S.; Taylor, H.; Ebdon, J.; Papageorgiou, G.; Pissarides, N.; Antoniou, K.; Christofi, G.; Dorlfinger, G.; et al. Simulation of daily discharge using the distributed model SWAT as a catchment management tool: Limnatis River case study. *Proc. SPIE Int. Soc. Opt. Eng.* **2013**. [CrossRef]

22. Tzoraki, O.; Cooper, D.; Kjeldsen, T.; Nikolaidis, N.P.; Froebrich, J.; Querner, E.; Gallart, F. Flood generation and classification of a semi-arid intermittent flow watershed: Evrotas river. *Int. J. River Basin Manag.* **2013**, *11*, 77–92. [CrossRef]

23. De Girolamo, A.; Gallart, F.; Pappagallo, G.; Santese, G.; Lo Porto, A. An eco-hydrological assessment method for temporary rivers. The Celone and Salsola rivers case study (SE, Italy). *Ann. Limnol. Int. J. Limnol.* **2015**, *51*, 1–10. [CrossRef]

24. D'Ambrosio, E.; De Girolamo, A.M.; Barca, E.; Ielpo, P.; Rulli, M.C. Characterising the hydrological regime of an ungauged temporary river system: A case study. *Environ. Sci. Pollut. Res.* **2017**, *24*, 13950–13966. [CrossRef]

25. Oueslati, O.; Girolamo, A.M.D.; Abouabdillah, A.; Kjeldsen, T.R.; Porto, A.L. Classifying the flow regimes of Mediterranean streams using multivariate analysis. *Hydrol. Process.* **2015**, *4682*, 4666–4682. [CrossRef]

26. Tzoraki, O. Operating Small Hydropower Plants in Greece under Intermittent Flow Uncertainty: The Case of Tsiknias River (Lesvos). *Challenges* **2020**, *11*, 17. [CrossRef]

27. Angela, D.; Krystalia, E.; Vaia, D.; Epaminondas, L.; Costas, A.; Panayiota, K.; Nikos, P.; Arsinoe, S.; Skevi, S.; Yioannis, B.; et al. Temporal and inter habitat variations of substratum, vegetation and substratum macroinvertebrates attributes across coastal wetland systems, North East Aegean, Greece. *Transit. Waters Bull.* **2008**, *2*, 1–16. [CrossRef]

28. Spyropoulou, A.; Spatharis, S.; Papantoniou, G.; Tsirtsis, G. Potential response to climate change of a semi-arid coastal ecosystem in eastern Mediterranean. *Hydrobiologia* **2013**, *705*, 87–99. [CrossRef]

29. *Natura 2000 European Environment Agency (EEA), The Natura 2000 Protected Areas Network*; EEA: Copenhagen, Denmark, 2000.

30. Borsi, S.; Ferrara, G.; Innocenti, F.; Mazzuoli, R. Geochronology and Petrology of Recent Volcanics in the Eastern Aegean Sea (West Anatolia and Lesvos Island). *Bullin Volcanol.* **1972**, *81*, 473–496. [CrossRef]

31. Pe-Piper, D.J.W. Spatial and temporal variation in Late Cenozoic volcanic rocks, Aegean Sea region. *Tecronophysics* **1989**, *169*, 113–134. [CrossRef]

32. Polatidou, M.; Tsirtsis, G.; Gaganis, P. Assessing nutrient dynamics in a small eastern mediterranean watershed. In Proceedings of the 13th International Conference on Environmental Science and Technology—CEST2013, Athens, Greece, 5–7 September 2013; pp. 5–7.

33. CLC 2000—Copernicus Land Monitoring Service. Available online: https://land.copernicus.eu/pan-european/corine-land-cover/clc-2000?tab=metadata (accessed on 16 February 2021).

34. Haines Young, R.; Weber, J.L. *Land Accounts for Europe 1990–2000: Towards Integrated Land and Ecosystem Accounting*; EEA: Copenhagen, Denmark, 2006; Volume 11.

35. Provatas, N. *Time and Spatial Analysis of Water Quality in Insular Basins: The Case of Tsiknias River, Lesvos*; University of the Aegean: Mytilene, Greece, 2015.

36. Zanis, P.; Kapsomenakis, I.; Philandras, C.; Douvis, K.; Nikolakis, D.; Kanellopoulou, E.; Zerefos, C.; Repapis, C. Analysis of an ensemble of present day and future regional climate simulations for Greece. *Int. J. Climatol.* **2009**, *29*, 1614–1633. [CrossRef]

37. Zanis, P.; Katragkou, E.; Ntogras, C.; Marougianni, G.; Tsikerdekis, A.; Feidas, H.; Anadranistakis, E.; Melas, D. Transient high-resolution regional climate simulation for Greece over the period 1960–2100: Evaluation and future projections. *Clim. Res.* **2015**, *64*, 123–140. [CrossRef]

38. Jacob, D.; Petersen, J.; Eggert, B.; Alias, A.; Christensen, O.B.; Bouwer, L.M.; Braun, A.; Colette, A.; Déqué, M.; Georgievski, G.; et al. EURO-CORDEX: New high-resolution climate change projections for European impact research. *Reg. Environ. Chang.* **2014**, *14*, 563–578. [CrossRef]
39. Zanis, P.; Akritidis, D.; Tsikerdekis, T.; Solomos, S.; Amiridis, V. Establishing a pilot regional climate change web application tool for end-users. In Proceedings of the GEO-CRADLE Workshop & Project Meeting, Limassol, Cyprus, 16–17 November 2016.
40. Willkofer, F.; Schmid, F.; Komischke, H.; Korck, J.; Braun, M.; Ludwig, R. Journal of Hydrology: Regional Studies The impact of bias correcting regional climate model results on hydrological indicators for Bavarian catchments. *J. Hydrol. Reg. Stud.* **2021**, *19*, 25–41. [CrossRef]
41. Gassman, P.W.; Reyes, M.R.; Green, C.H.; Arnold, J.G. The Soil And Water Assessment Tool: Historical Development, Applications, and Future Research Directions. *Am. Soc. Agric. Biol. Eng.* **2007**, *50*, 1211–1250. [CrossRef]
42. USDA-SCS. *National Engineering Handbook*; USDA-SCS: Washington, DC, USA, 1972.
43. Winchell, M.; Srinivasan, R.; Di Luzio, M. *ArcSWAT 2.3. 4 Interface for SWAT2012*; ArcSWAT: Temple, TX, USA, 2013.
44. Abbaspour, K.C.; Johnson, C.A.; Van Genuchten, M.T. Estimating Uncertain Flow and Transport Parameters Using a Sequential Uncertainty Fitting Procedure. *Vadose Zone J.* **2004**, *3*, 1340–1352. [CrossRef]
45. Abbaspour, K.C.; Yang, J.; Maximov, I.; Siber, R.; Bogner, K.; Mieleitner, J.; Zobrist, J.; Srinivasan, R. Modelling hydrology and water quality in the pre-alpine/alpine Thur watershed using SWAT. *J. Hydrol.* **2007**, *333*, 413–430. [CrossRef]
46. Abbaspour, K.C.; Rouholahnejad, E.; Vaghefi, S.; Srinivasan, R.; Yang, H.; Kløve, B. A continental-scale hydrology and water quality model for Europe: Calibration and uncertainty of a high-resolution large-scale SWAT model. *J. Hydrol.* **2015**, *524*, 733–752. [CrossRef]
47. Karavitis, C.A.; Kerkides, P.; Karavitis, C.A.; Iwra, M.; Collins, F. Estimation of the Water Resources Potential in the Island System of the Aegean Archipelago, Greece Estimation of the Water Resources Potential in the Island System of the Aegean Archipelago, Greece. *Water Int.* **2009**. [CrossRef]
48. Varvara, M.; Christos, V.; Ourania, T.; Kostas, K. Using Remote Sensing technology with SWAT hydrological modeling to estimate soil moisture of an insular basin. In Proceedings of the IWA Balkan Young Water Professionals Conference 2015, Thessaloniki, Greece, 10–12 May 2015. Available online: http://bywp2015.gr (accessed on 22 January 2021).
49. Simha, P.; Mutiara, Z.Z.; Gaganis, P. Vulnerability assessment of water resources and adaptive management approach for Lesvos Island, Greece. *Sustain. Water Resour. Manag.* **2017**. [CrossRef]
50. Bormann, H.; Brito, M.M.D.; Charchousi, D.; Chatzistratis, D.; Korali, A.; Krauzig, N.; Meier, J.; Meliadou, V.; Meinhardt, M. Impact of Hydrological Modellers' Decisions and Attitude on the Performance of a Calibrated Conceptual Catchment Model: Results from a "Modelling Contest". *Hydrology* **2018**, *5*, 64. [CrossRef]
51. Gupta, R.D.; Kundu, D. Generalized exponential distributions. *Austral. Newzeal. J. Stat.* **1999**, *41*, 173–188. [CrossRef]
52. Nash, J.E.; Sutcliffe, J.V. River flow forecasting through conceptual models part I—A discussion of principles. *J. Hydrol.* **1970**, *10*, 282–290. [CrossRef]
53. Moriasi, D.N.; Gitau, M.W.; Pai, N.; Daggupati, P. Hydrologic and Water Quality Models: Performance Measures And Evaluation Criteria. *Am. Soc. Agric. Biol. Eng.* **2015**, *58*, 1763–1785. [CrossRef]
54. Moriasi, D.N.; Arnold, J.G.; Liew, M.W.V.; Bingner, R.L.; Harmel, R.D.; Veith, T.L. Model Evaluation Guidelines For Systematic Quatification of accuracy in Watershed Simulations. *Am. Soc. Agric. Biol. Eng.* **2007**, *50*, 885–900.
55. Gallart, F.; Latron, J.; Llorens, P.; Cid, N.; Prat, N.; Rieradevall, M. *Deliverable 9: The TREHS Manual*; Life TRivers: Barcelona, Spain, 2015.
56. Gallart, F.; Prat, N.; Garc, E.M. A novel approach to analysing the regimes of temporary streams in relation to their controls on the composition and structure of aquatic biota. *Hydrol. Earth Syst. Sci.* **2012**, *16*, 3165–3182. [CrossRef]
57. Gallart, F.; Llorens, P.; Latron, J.; Cid, N.; Rieradevall, M.; Prat, N. Validating alternative methodologies to estimate the regime of temporary rivers when fl ow data are unavailable. *Sci. Total Environ.* **2016**, *565*, 1001–1010. [CrossRef] [PubMed]
58. Gallart, F.; Prat, N.; García-Roger, E.M.; Latron, J.; Rieradevall, M.; Llorens, P.; Barberá, G.G.; Brito, D.; De Girolamo, A.M.; Lo Porto, A.; et al. Developing a novel approach to analyse the regimes of temporary streams and their controls on aquatic biota. *Hydrol. Earth Syst. Sci. Discuss.* **2011**, *8*, 9637–9673. [CrossRef]
59. Foster, H. Duration curves. *Trans (ASCE)* **1934**, *99*, 1213–1267.
60. Stanzel, P.; Kling, H. From ENSEMBLES to CORDEX: Evolving climate change projections for Upper Danube River flow. *J. Hydrol.* **2018**. [CrossRef]
61. Lionello, P.; Scarascia, L. The relation of climate extremes with global warming in the Mediterranean region and its north versus south contrast. *Reg. Environ. Chang.* **2020**, *20*. [CrossRef]
62. De Girolamo, A.M.; Lo Porto, A.; Pappagallo, G.; Gallart, F. Assessing flow regime alterations in a temporary river—The River Celone case study. *J. Hydrol. Hydromech.* **2015**, *63*, 263–272. [CrossRef]
63. Skoulikaris, C.; Makris, C.; Katirtzidou, M.; Baltikas, V.; Krestenitis, Y. Assessing the Vulnerability of a Deltaic Environment due to Climate Change Impact on Surface and Coastal Waters: The Case of Nestos River (Greece). *Environ. Model. Assess.* **2021**, 1–28. [CrossRef]
64. De Girolamo, A.M.; Bouraoui, F.; Buffagni, A.; Pappagallo, G.; Lo Porto, A. Hydrology under climate change in a temporary river system: Potential impact on water balance and flow regime. *River Res. Appl.* **2017**, *33*, 1219–1232. [CrossRef]

65. Kirkby, M.J.; Gallart, F.; Kjeldsen, T.R.; Irvine, B.J.; Froebrich, J.; Lo Porto, A.; De Girolamo, A. Classifying low flow hydrological regimes at a regional scale. *Hydrol. Earth Syst. Sci.* **2011**, *15*, 3741–3750. [CrossRef]
66. Alderlieste, M.A.A.; Van Lanen, H.A.J.; Wanders, N. Future low flows and hydrological drought: How certain are these for Europe? *IAHS-AISH Proc. Rep.* **2014**, *363*, 60–65.
67. Girolamo, A.M.D.; Barca, E.; Pappagallo, G.; Porto, A.L. Simulating ecologically relevant hydrological indicators in a temporary river system. *Agric. Water Manag.* **2017**, *180*, 194–204. [CrossRef]
68. Skoulikidis, N.T.; Sabater, S.; Datry, T.; Morais, M.M.; Buffagni, A.; Dörflinger, G.; Zogaris, S.; del Mar Sánchez-Montoya, M.; Bonada, N.; Kalogianni, E.; et al. Non-perennial Mediterranean rivers in Europe: Status, pressures, and challenges for research and management. *Sci. Total Environ.* **2017**, *577*, 1–18. [CrossRef]
69. Vlach, V.; Ledvinka, O.; Matouskova, M. Changing Low Flow and Streamflow Drought Seasonality in Central European Headwaters. *Water* **2020**, *12*, 3575. [CrossRef]
70. Njue, N.; Kroese, J.S.; Gräf, J.; Jacobs, S.R.; Weeser, B.; Breuer, L.; Ru, M.C. Science of the Total Environment Citizen science in hydrological monitoring and ecosystem services management: State of the art and future prospects. *Sci. Total Environ.* **2019**, *693*. [CrossRef] [PubMed]

Article

Evaluation and Calibration of Alternative Methods for Estimating Reference Evapotranspiration in the Senegal River Basin

Papa Malick Ndiaye [1], Ansoumana Bodian [1,*], Lamine Diop [2], Abdoulaye Deme [3], Alain Dezetter [4] and Koffi Djaman [5]

[1] Laboratoire Leïdi Dynamique des Territoires et Développement, Université Gaston Berger (UGB), BP 234 Saint Louis, Senegal; ndiaye.papa-malick@ugb.edu.sn

[2] UFR S2ATA Sciences Agronomiques, de l'Aquaculture et des Technologies Alimentaires, Université Gaston Berger, BP 234 Saint-Louis, Senegal; lamine.diop@ugb.edu.sn

[3] Laboratoire LSAO "Laboratoire des Sciences de l'Atmosphère et de l'Océan", Université Gaston Berger (UGB), BP 234 Saint-Louis, Senegal; abdoulaye.deme@ugb.edu.sn

[4] HydroSciences Montpellier, University of Montpellier, IRD, CNRS, CC 057, 163 rue Auguste Broussonnet, 34090 Montpellier, France; Alain.Dezetter@ird.fr

[5] Department of Plant and Environmental Sciences, New Mexico State University, Agricultural Science Center at Farmington, P.O. Box 1018, Farmington, NM 87499, USA; kdjaman@nmsu.edu

* Correspondence: bodianansoumana@gmail.com or ansoumana.bodian@ugb.edu.sn; Tel.: +221-77-811-7553

Received: 3 April 2020; Accepted: 24 April 2020; Published: 28 April 2020

Abstract: Reference evapotranspiration (ET_0) is a key element of the water cycle in tropical areas for the planning and management of water resources, hydrological modeling, and irrigation management. The objective of this research is to assess twenty methods in computing ET_0 in the Senegal River Basin and to calibrate and validate the best methods that integrate fewer climate variables. The performance of alternative methods compared to the Penman Monteith (FAO56-PM) model is evaluated using the coefficient of determination (R^2), normalized root mean square error (NRMSE), percentage of bias (PBIAS), and the Kling–Gupta Efficiency (KGE). The most robust methods integrating fewer climate variables were calibrated and validated and the results show that Trabert, Valiantzas 2, Valiantzas 3, and Hargreaves and Samani models are, respectively, the most robust for ET_0 estimation. The calibration improves the estimates of reference evapotranspiration compared to original models. It improved the performance of these models with an increase in KGE values of 45%, 32%, 29%, and 19% for Trabert, Valiantzas 2, Valiantzas 3, and Hargreaves and Samani models, respectively. From a spatial point of view, the calibrated models of Trabert and Valiantzas 2 are robust in all the climatic zones of the Senegal River Basin, whereas, those of Valiantzas 3 and Hargreaves and Samani are more efficient in the Guinean zone. This study provides information on the choice of a model for estimating evapotranspiration in the Senegal River Basin.

Keywords: reference evapotranspiration; FAO56-PM; alternative methods; calibration/validation; Senegal River basin

---

1. Introduction

Evapotranspiration (ET) is an essential component for the planning and management of water resources [1–5], irrigation programming [6], drought studies [7], and climate change [8,9]. In addition, when combined with precipitation, evapotranspiration can be a drought index and a classification tool for climates [7,10]. In the agricultural field, evapotranspiration is an essential parameter for the management of water resources and the optimization of irrigation at the plot scale [11]. Indeed, it is used for the estimation of crop water requirements [12]. Evapotranspiration is also a climate synthesis and therefore an indicator of climate change [13,14]. In fact, the evapotranspiration process is controlled by climate variables (temperature, solar radiation, relative humidity, wind), edaphic variables (soil type), and physiological factors [15].

Reference evapotranspiration (ET_0) is the estimation of the evapotranspiration from a hypothetical grass reference actively growing, completely shading the ground and not short of water with an assumed crop height of 0.12 m, a fixed surface resistance of 70 s/m, and an albedo of 0.23 [16,17]. ET_0 can be determined by in situ measurements (lysimeter, pan, atmometer, scintillometer, Eddy covariance) or computed from weather data [18]. Direct measurement is indicated for evapotranspiration estimation [19–21]. However, instruments are expensive and difficult to maintain [2,18]. Therefore, several authors [22–27] have developed evapotranspiration estimation methods. Among these methods, Penman Monteith (FAO56-PM) is recommended as a reference method [18]. However, the number of climate variables (temperature, radiation, wind speed, and relative humidity) that it integrates constrains its use in developing countries where access to climate data is difficult [28–30].

In this context, alternative methods integrating fewer climate variables are used [22,23,25,27,31–33]. These alternative methods are classified into four categories according to climate variables that integrate [3] (i) aerodynamic [22,31,33–35], (ii) temperature-based [25,26,36–38], (iii) radiation-based [2,23,31,39,40], (iv) and combinatory methods [3,24,27]. These different methods have been developed in specific contexts. Therefore, they must be calibrated and validated for improving their performance and adapt them to other climate conditions [1,29]. In this regard, different methods of estimating ET_0 have been evaluated and calibrated under various climate conditions around the world [3,12,29,41–44]. In the Senegal River Valley, Djaman et al. [29] evaluated 15 evapotranspiration estimation methods. They have shown that the models of Valiantzas, Trabert, Romanenko, Schendel, and Mahinger are more robust in this area. In another study, Djaman et al. [45] evaluated and calibrated six methods for estimating ET_0 in the Senegal River Delta. Their results show that the Valiantzas 2 method is more effective among the six methods evaluated. However, these two studies are limited to specific areas of the Senegal River Basin. To our knowledge, no study has been interested in the evaluation and calibration/validation of alternative methods of estimating ET_0 at the Senegal River Basin. Thus, the objective of this study is to evaluate 20 methods for estimating reference evapotranspiration and to calibrate and validate the best methods integrating fewer climate variables in order to adapt them to the context of the Senegal River Basin.

2. Materials and Methods

2.1. Study Area

The Senegal River Basin covers an area of 300,000 km^2 [46] and extends over four countries: Guinea, Mali, Senegal, and Mauritania (Figure 1). According to the latitudinal distribution of precipitation, Dione [47] identified four main climatic zones: Guinean (average annual rainfall: P > 1500 mm), South Sudanian (1000 < P < 1500 mm), North Sudanian (500 < P < 1000 mm), and Sahelian (P < 500 mm). The Senegal River average annual flow at Bakel station (Figure 1) over the period 1950–2014 was 600 m^3/s, an average annual volume of 18 billion m^3. Bakel is considered as the reference station of the Senegal river basin because it is located downstream from the three Senegal River tributaries: Bafing, the Bakoye, and the Faleme. Senegal River water resources are used for hydroelectricity production, navigation, drinking water supply and irrigated agriculture [28]. The potential irrigable land at the Senegal River Basin is estimated at 408,900 hectares with an irrigated area of 21,2937 hectares [48]. The percentage of exploited potential irrigable land varies from 45 to 68% depending on the country (Figure 1). Several hydraulic infrastructures have been built by the Organization for the Development of the Senegal River (in French, Organisation pour la Mise en Valeur du fleuve Sénégal, OMVS): Diama in 1986, Manantali in 1988, and Felou in 2013. The Diama dam's role is to stop the saltwater intrusion and to allow the development of irrigation in the Senegal River Valley and the Delta. Manantali, on the Bafing tributary, is a multifunctional dam with a capacity of 11 billion m^3 of water, thus allowing an electrical production of 800 GWh/year and an irrigation capacity of 255,000 ha. Felou is a run-of-river dam with an electricity production capacity of 350 GWh/year. Several other dams are planned (Figure 1) to increase the production of hydroelectricity and to regulate the Faleme and Bakoye tributaries.

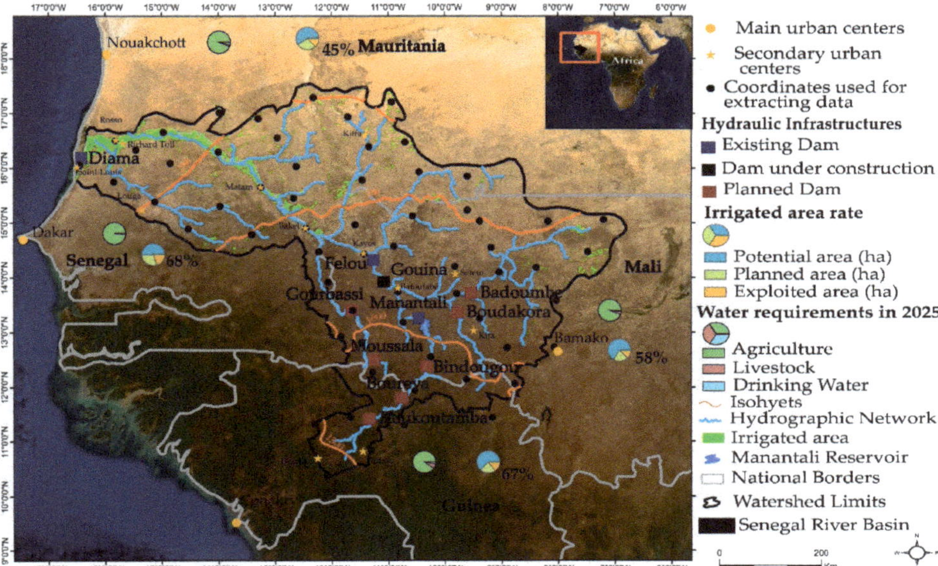

Figure 1. Senegal River Basin, stations used for the extraction of climatic variables, hydraulic infrastructures, water uses, and needs by sector of activity.

2.2. Data

In West Africa, climate data managed by national meteorological services are difficult for researchers to access due to their high cost of acquisition [30,49]. In addition, the low density of the observation network poses a problem of representativeness of these data at the scale of the watershed. However, the large-scale study of evapotranspiration requires several measurement points due to the heterogeneity of the landscapes and the variation in energy transfer processes [50]. Therefore, in this study, both reanalysis and observation data from the NASA Langley Research Center (LaRC) POWER Project funded through the NASA Earth Science/Applied Science Program (https://power.larc.nasa.gov, accessed on 20 December 2018) were used as an alternative to the observed weather station data that are inaccessible or scattered. These data have the advantage of having a spatial and temporal coverage on a global scale [51–55] and provide the climate variables necessary for the estimation of reference evapotranspiration [48,53]. For the extraction of reanalysis data, the coordinates of rainfall stations (Figure 1) from the OMVS database [29] were used. The climate variables extracted on a daily basis over the period 1984–2017 are: temperature (max and min), relative humidity, wind speed and solar radiation. A summary of these variables according to climate zones is given in Table 1.

Table 1. Average values of climatic variables.

Climate Zone *	u2 (m s^{-1})	Tmax (°C)	Tmin (°C)	Rh (%)	Rs (MJ m^{-2})	ET$_0$ (mm day^{-1})
Guinean	1.70	30.42	21.18	67.65	19.65	4.54
Sudanian	2.22	34.88	22.45	42.03	20.72	6.30
Sahelian	3.00	37.12	22.95	29.00	21.29	8.01

Read: u2 wind speed, Tmax maximum temperature, Tmin minimum temperature, Rh relative humidity, Rs solar radiation, ET$_0$ reference evapotranspiration, * according to Dione's breakdown [47], the basin is subdivided into four climate zones (Guinean, South Sudanese, North Sudanian, and Sahelian), but in this study, for a better readability of the results, the subdivisions of the Sudanian zone were not taken into account. Thus, the zones considered were Guinean, Sudanese, and Sahelian.

2.3. Method

The methodology was organized into the following three steps: (i) calculation of the reference evapotranspiration by the FAO56-PM method and by 20 alternative methods, (ii) evaluation of the performance of these alternative methods compared to the FAO56-PM, and (iii) calibration and validation of the best methods integrating fewer climate variables.

2.3.1. Estimation of Reference Evapotranspiration

FAO56-PM (1) and twenty alternative methods (2–21) were used to estimate the reference evapotranspiration. FAO56-PM consider the following characteristics defined by Allen et al. [18]: reference surface characterized by short and green vegetation (grass for this study), adequately supplied in water, uniform height (0.12 m), albedo of 0.23, and resistance surface of 70 m/s. This method is recommended as a reference method for the estimation of ET$_0$ without adjustment or integration of parameters [18]. Its formulation is as follows:

$$ET_{0FAO56-PM} = \frac{0.408\Delta(Rn - G) + \frac{\gamma Cn}{T+273.3}u2\,(es - ea)}{\Delta + \gamma(1 + Cdu2)} \tag{1}$$

where ET$_{0FAO56-PM}$ is the reference evapotranspiration (mm/day); Rn: net radiation on the crop surface (MJ·m^{-2}d^{-1}); G is the heat flux density of the soil (MJ·m^{-2}d^{-1}), which is ignored on a daily scale; T is the average daily air temperature at a height of 2 m (°C); Cn and Cd are constant values, which change according to the scale of time used (on a daily scale Cn and Cd are 900 and 0.34 respectively); u2 is the wind speed at a height of 2 m (ms^{-1}); es is the saturated vapor pressure (kPa); ea is real vapor pressure (kPa); (es—ea) is the saturation deficit (kPa); Δ is the vapor pressure slope curve (kPa°C^{-1}); and γ is the psychrometric constant (kPa°C^{-1}).

The alternative methods are classified into the following four categories (Table 2): aerodynamic, radiation-based, temperature-based, and combinatory methods. The choice of these methods is justified by their frequent use, the simplicity of their implementation, and their performance under different climate conditions.

Table 2. Characteristics of the 20 alternative methods used.

Categories	References	Formula	Abbreviation	
Aerodynamic	Dalton [21]	$ET_0 = (0.3648 + 0.07223u2)(es - ea)$	DN	(2)
	Trabert [30]	$ET_0 = 0.3075 \sqrt{u2}(es - ea)$	TRB	(3)
	Penman [31]	$ET_0 = 0.35(1 + 0.24u2)(es - ea)$	PNM	(4)
	Rohwer [33]	$ET_0 = 0.44(1 + 0.27u2)(es - ea)$	RW	(5)
	Mahringer [34]	$ET_0 = 0.15072 \sqrt{3.6}u2(es - ea)$	MHR	(6)
Temperature	Hargreaves [24]	$ET_0 = 0.0135 \times 0.408Rs(T + 17.8)$	HG	(7)
	Hargreaves and Samani [25]	$ET_0 = 0.408 \times 0.0023(T + 17.8)(Tmax - Tmin)^{0.5}Ra$	HS	(8)
	Trajkovic [35]	$ET_0 = 0.408 \times 0.0023(T + 17.8)(Tmax - Tmin)^{0.424}Ra$	TRA	(9)
	Droogers and Allen [36]	$ET_0 = 0.408 \times 0.0025(T + 16.8)(Tmax - Tmin)^{0.5}Ra$	DA	(10)
	Heydari and Heydari [37]	$ET_0 = 0.0023Ra(T + 9.519)(Tmax - Tmin)^{0.611}$	HH	(11)
Radiation	Makkink [22]	$ET_0 = 0.61 \frac{\Delta}{\Delta+\gamma} \times \frac{Rs}{\lambda} - 0.012$	MK	(12)
	Jensen and Haise [38]	$ET_0 = 0.025(T - 3)Rs$	JH	(13)
	Priestley and Taylor [39]	$ET_0 = \alpha\frac{\Delta}{\Delta+\gamma} \times \frac{Rn}{\lambda}$	PT	(14)
	Abtew [32]	$ET_0 = 0.53\frac{Rs}{\lambda}$	AB	(15)
	Oudin [2]	$ET_0 = Rs \times \frac{T+5}{100}$	OD	(16)
Combinatory	Penman [23]	$ET_0 = \left[\frac{\Delta}{\Delta+\gamma}(Rn - G) + \frac{\gamma}{\Delta+\gamma} \times 6.43(1 + 0.053u2)(es - ea)\right]/\lambda$	PNM	(17)
	Doorenboss and Pruitt [40]	$ET_0 = \left[\frac{\Delta}{\Delta+\gamma}(Rn - G) + 2.7\frac{\gamma}{\gamma+\Delta}(1 + 0.864u2)(es - ea)\right]/\lambda$	DP	(18)
	Valiantzas [26]	$ET_0 = 0.0393Rs \sqrt{T + 9.5} - 0.19Rs^{0.6} \varphi^{0.15} + 0.048(T + 20)\left(1 - \frac{Hr}{100}\right)u2^{0.7}$	Val 1	(19)
	Valiantzas [26]	$ET_0 = 0.0393Rs \sqrt{T + 9.5} - 0.19Rs^{0.6} \varphi^{0.15} + 0.078(T + 20)\left(1 - \frac{Hr}{100}\right)$	Val 2	(20)
	Valiantzas [26]	$ET_0 = 0.0393Rs \sqrt{T + 9.5} - 0.19Rs^{0.6} \varphi^{0.15} + 0,0061(T + 20)(1.12T - Tmin - 2)^{0.7}$	Val 3	(21)

Read: ET_0 reference evapotranspiration (mm); u2 represents the wind speed measured at 2 m from the ground (ms^{-1}); (es—ea) saturation deficit (kPa); T is the average temperature (°C); Tmax—maximum temperature (°C); Tmin—minimum temperature (°C); Ra is the extraterrestrial radiation (MJ m^{-2}d^{-1}); Δ is the saturating vapor pressure curve (kPa°C^{-1}); γ is the psychrometric constant (kPa°C^{-1}); λ is the latent heat of vaporization (MJ m^{-2}d^{-1}); Rs is the short wavelength solar radiation (MJ m^{-2}d^{-1}); Rn is the radiation net (MJ m^{-2}d^{-1}) and α is a constant value (1.26 for humid areas and 1.74 for semi-arid areas). These coefficients are considered to be constant for a given region [41]. φ represents the latitude of the station in radian degree, and λ is the latent heat of vaporization (MJ.m^{-2}d^{-1}).

2.3.2. Performance of the Alternative Methods

The performance of alternative methods for estimating ET_0 is evaluated using the coefficient of determination (R^2), the normalized root mean square error (NRMSE), the percentage of bias (PBIAS), and (iv) the Kling–Gupta Efficiency (KGE) [56]. R^2 provides information on the degree of agreement, the NRMSE estimates the average deviation and PBIAS gives the underestimation/overestimation of ET_0 by alternative methods. KGE combines both the correlation coefficient (r), the biases (β), and the variability (γ). The formulation of these different criteria, their amplitude of variation and their optimal value are given in Table 3.

Table 3. Range and optimum value of the evaluation criteria.

Criteria.	Formula	Range	Optimum Value	
R^2	$\dfrac{\sum_{i=1}^{n}\left(ET_{0alt}-\overline{ET_{0FAO56-PM}}\right)^2}{\sum_{i=1}^{n}\left(ET_{0FAO56-PM}-\overline{ET_{0FAO56-PM}}\right)^2}$	0 to 1	1	(22)
NMRSE	$\dfrac{\sqrt{\frac{1}{n}\sum_{i=1}^{n}\left(ET_{oalt}-ET_{0FAO56-PM}\right)^2}}{\frac{1}{n}\sum_{i=1}^{n}\left(ET_{oalt}\right)}$	0 to $+\infty$	0	(23)
PBIAS	$\left[\dfrac{\frac{1}{n}\sum_{i=1}^{n}\left(ET_{oalt}-ET_{0FAO56-PM}\right)^2}{\frac{1}{n}\sum_{i=1}^{n}\left(ET_{oalt}\right)}\right]*100$	$-\infty$ to $+\infty$	0	(24)
KGE	$1-\sqrt{(r-1)^2+(\beta-1)^2+(\alpha-1)^2}$	$-\infty$ to 1	1	(25)

R^2—coefficient of determination; ET_{0alt}—evapotranspiration estimated by an alternative method; $ET_{0FAO56-PM}$—evapotranspiration estimated by the FAO56-PM method; NMRSE—normalized mean error between alternative methods and that of FAO56-PM; PBIAS—percentage of biases between methods (negative values represent an underestimation and positive ones an overestimation). KGE is the Kling–Gupta Efficiency coefficient; it is made up of three variables: r—the correlation coefficient between the alternative methods evaluated and that of FAO56-PM, α—the variability, and β—the gaps that exist between the alternative methods evaluated and that of FAO56-PM.

2.3.3. Calibration and Validation

Two criteria are used to choose the methods to be calibrated—the performance of the method to estimate ET_0 and the number of climate variables that it integrates. The methods integrating only two or three climate variables are preferred for calibration/validation. The calibration consists of changing the constant values of the methods in order to increase their performance [57]. The objective is to optimize the value of the KGE and reduce the errors obtained during the evaluation. For this, the series is divided into two parts as recommended by Xu and Singh [1]: 2/3 of the series (1984–2005) for calibration and 1/3 (2006–2017) for validation. The calibration is done by applying the generalized method of gradient reduction [58]. For each method, a constant value is changed to optimize the KGE and reduce the NRMSE by using iteration method. R^2, NRMSE, PBIAS, and KGE, as well as the Taylor diagram [59], are used to assess the performance of the methods after calibration/validation.

3. Results and Discussion

3.1. Performance of the Twenty Methods

Figure 2 shows the performance of the twenty alternative methods according to the four criteria selected. The combinatory methods of Valiantzas 1 (Val 1), Doorenboss and Pruitt (DP), and Penman (PN) are more robust for estimating reference evapotranspiration. Indeed, they have high coefficients of determination and KGE: Val 1 ($R^2 = 0.96$, KGE = 0.93), DP ($R^2 = 0.90$, KGE = 0.85), and PN ($R^2 = 0.96$, KGE = 0.66). The errors of estimation of ET_0 by these methods are low with NRMSE values of 0.06, 0.11, and 0.18 for Val 1, DP, and PN, respectively. PBIAS analysis shows that Val 1 and DP methods slightly underestimate ET_0 (PBIAS of -2.23 for Val 1 and -9.63 for DP). In contrast, the PN method overestimates ET_0 by 16.13%. The method of Val 2 has values of R^2 and KGE of 0.67 and 0.55 and that of Val 3 of 0.47 and 0.42. However, compared to the FAO56-PM method, Val 2 underestimates ET_0 by 48.9% and Val 3 by 61.9%.

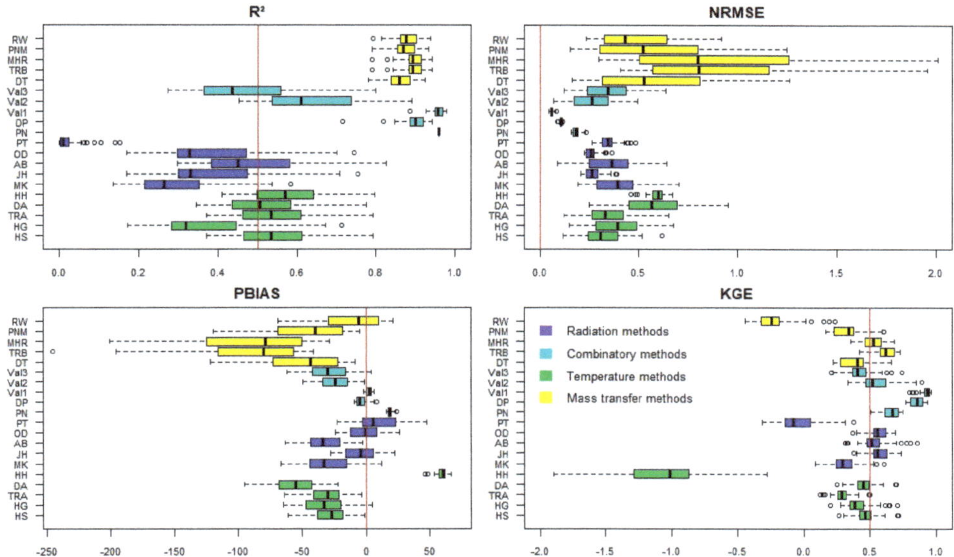

Figure 2. Performance of the methods according to the selected evaluation criteria. The red line in each figure represents the threshold values of each evaluation criterion.

Figure 3 shows the spatial distribution of KGE values. It shows that the combinatory models of Val 1, DP, and PN are robust in all Senegal River basin climate zones. They have KGE values that vary from 0.51 to 0.97 over the entire basin. The performance of these methods is explained by the fact that they integrate the same climate variables as those of FAO56-PM. Among the methods integrating fewer climate variables, Trabert, Val 2, Val 3, HS, and JH are most robust (Figures 2 and 3). The performance of these methods varies depending on the climate zones of the basin. Indeed, Trabert's aerodynamic method is more accurate in the Sudanian and Sahelian zones with KGE values varying from 0.69 to 0.73. Similar results were obtained by Djaman et al. [29] in the Senegal River Valley. The performance of aerodynamic methods was also noted by [3] in Iran, Singh and Xu [60] in northwestern Ontario (Canada), and Djaman et al. [61] in Tanzania and Kenya. However, Ndiaye et al. [62] have shown that aerodynamic methods perform poorly in Burkina Faso. This difference could be explained by the constant values of these methods that do not adapt to all climate conditions. The performance of aerodynamic methods is explained by the fact that wind speed and temperature play an important role in evapotranspiration in arid and semi-arid environments [63].

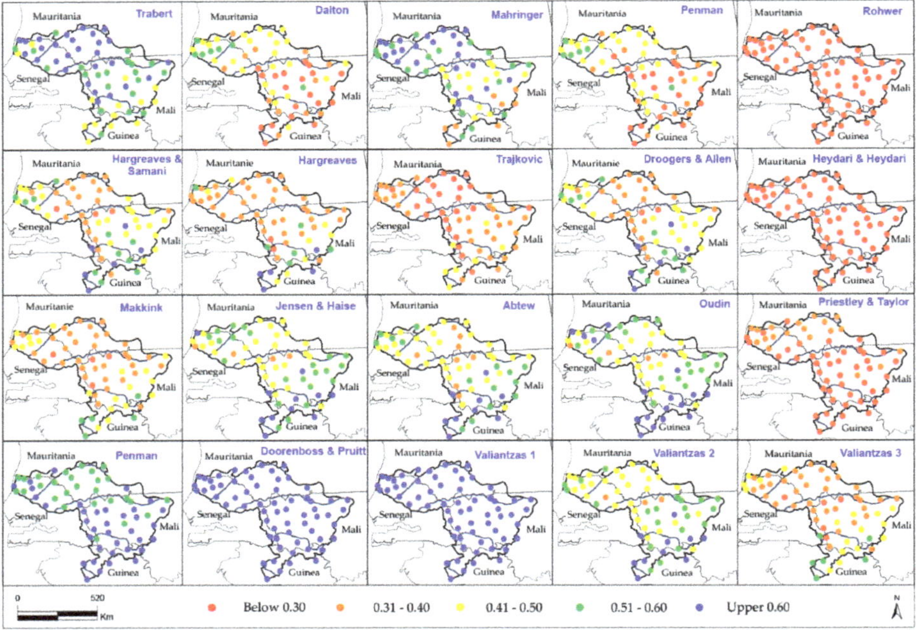

Figure 3. Spatial repartition of Kling–Gupta Efficiency (KGE) values.

In the Guinean zone, the combinatory (Val 2, Val 3) and temperature-based (HS) methods give the best results for estimating ET_0. Indeed, the values of KGE (Figures 2 and 3) of Val 2 vary from 0.82 to 0.90, those of Val 3 from 0.53 to 0.74, and the KGE of the HS model vary from 0.53 to 0.74. Errors in estimating ET_0 by these methods are also small in the Guinean area. These results are similar with those of Tabari [28] who showed the robustness of radiation and temperature-based methods in a humid climate in Iran. The performance of combinatory and temperature-based methods in the Guinean area of the basin is explained by the fact that they integrate solar radiation and temperature which have more impact on ET_0 in humid climates [64,65]. According to these authors, in humid regions, the air is close to saturation and evapotranspiration is strongly influenced by the available energy (temperature and radiation).

The evaluation of the twenty alternative methods for estimating reference evapotranspiration shows overall that the combinatory methods of Doorenboss and Pruitt, Penman, Valiantzas 1, 2, the aerodynamics models of Trabert and Mahinger, the based-on temperature of Hargreaves and Samani, and Jensen and Haise's radiation-based models present the best estimates of reference evapotranspiration. The methods of Rohwer (aerodynamics), Heydari and Heydari (based on temperature), and Priestley and Taylor (based on radiation) are less robust among the 20 methods. Based on their performance and the reduced number of climate variables they integrate, the methods of TRB, Val 2, Val 3, HS, and JH are selected for calibration.

3.2. Calibration and Validation of the Best Methods

Table 4 gives the best methods for estimating ET_0 before and after calibration, and Figures 4 and 5 give the performance of these methods according to the previously selected evaluation criteria.

Table 4. Best reference evapotranspiration (ET$_0$) estimation methods before and after calibration.

References	Before Calibration	After Calibration
Trabert [30]	$ET0_{TR} = 0.3075\ \sqrt{u2}(es - ea)$	$ET0_{TRcal} = 2.770\ \sqrt{u2}(es - ea)$
Valiantzas [26]	$ET0_{val2} = 0.0393Rs \times \sqrt{T+9.5} - 0.19Rs^{0.6}\ \varphi^{0.15} + 0.078(T+20)\left(1 - \frac{Hr}{100}\right)$	$ET0_{val2cal} = 0.027Rs \times \sqrt{T+9.5} - 0.19Rs^{0.6}\ \varphi^{0.15} + 0.159(T+20)\left(1 - \frac{Hr}{100}\right)$
Valiantzas [26]	$ET0_{val3} = 0.0393Rs \times \sqrt{T+9.5} - 0.19Rs^{0.6}\ \varphi^{0.15} + 0.0061(T+20)(1.12T - Tmin - 2)^{0.7}$	$ET0_{val3cal} = 0.026Rs \times \sqrt{T+9.5} - 0.19Rs^{0.6}\ \varphi^{0.15} + 0.018(T+20)(1.12T - Tmin - 2)^{0.7}$
Jensen and Haise [38]	$ET0_{JH} = 0.025(T - 3)Rs$	$ET0_{JHcal} = 0.027(T - 3)Rs$
Hargreaves and Samani [25]	$ET0_{HS} = 0.408 \times 0.0023(T + 17.8)(Tmax - Tmin)^{0.5}Ra$	$ET0_{HScal} = 0.408 \times 0.0031(T + 17.8)(Tmax - Tmin)^{0.5}Ra$

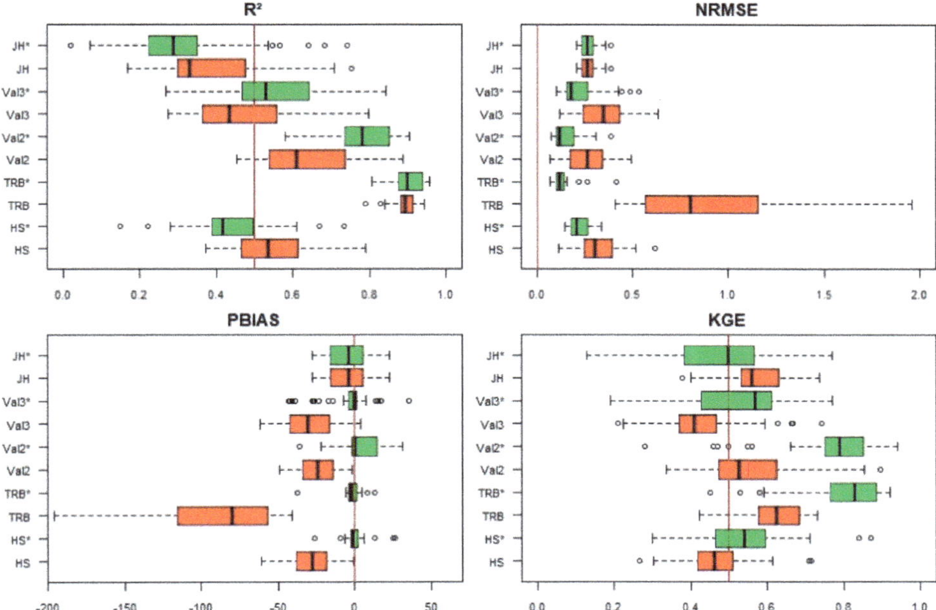

Figure 4. Method performance before and after calibration: (red color: before calibration; green color: After calibration).

The calibration/validation improved the performance of the TRB, Val 2, Val 3, and HS methods. In fact, it increased the KGE of Trabert's method by 32% and reduced estimation errors by 97%. However, a deterioration in the performance of the JH method is noted after calibration which is different from Irmak et al. [63], who reported that calibration of the JH method gives better results in humid climates. This deterioration could be explained by the poor performance of this method in semi-arid climates

The KGE values of the Val 2, Val 3, and HS methods are improved by 45%, 29%, and 19%, respectively, after calibration. On the other hand, the KGE of the JH method is degraded by 16% after calibration. Analysis of the results of the calibration/validation shows that the TRB and Val 2 methods always remain robust over the entire basin, while those of Val 3 and HS always remain more robust in the Guinean field. This is explained by the role of wind on the ET$_0$ in arid environments and that of temperature and radiation in humid climates.

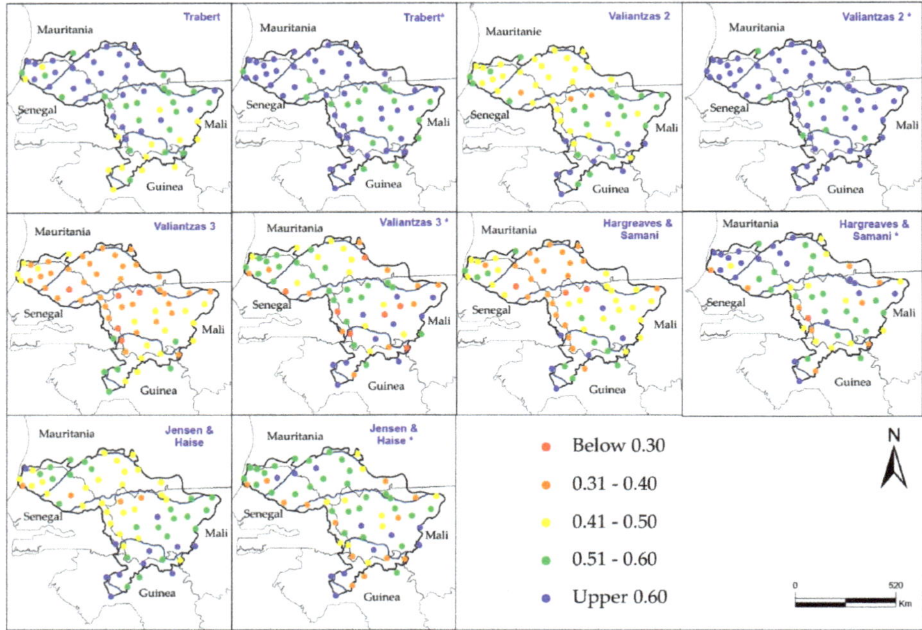

Figure 5. KGE values obtained after calibration of the best methods over the period 1984–2005. (* calibrated methods).

Data for the period 2006–2017 are used for the validation of the calibrated methods. Figure 6 gives the Taylor diagram which compares the results of the five validated methods and the FAO56-PM one as a function of climate zones, and Figure 7 gives the distribution of PBIAS. Results show that the methods of TRB and Val 2 always give the best estimates of ET_0. They have the same amplitude of variation as the FAO56-PM method with high correlation coefficients between 0.93 and 0.95. The spatial distribution of the percentages of bias allows us to note that the Trabert method globally underestimates evapotranspiration from 1.1 to 37%. The most significant underestimates are noted in the Guinean zone where the Trabert method is less efficient. The Val 2 method overestimates the ET_0 by 0.3%–31%, and the Val 3 methods, HS, and JH underestimate the reference evapotranspiration. These results are corroborated by those of Djaman et al. [29], who showed that the Trabert method underestimates the ET_0 by 25% at the Ndiaye station in the Senegal river delta and by 6% at the Fanaye station in the Senegal river valley. In another study, Djaman et al. [45] showed that the Val 2 method underestimates the ET_0 by-.13 mm in the Senegal delta. Ndiaye et al. [62] also showed that Trabert's method underestimates ET_0 in Burkina Faso.

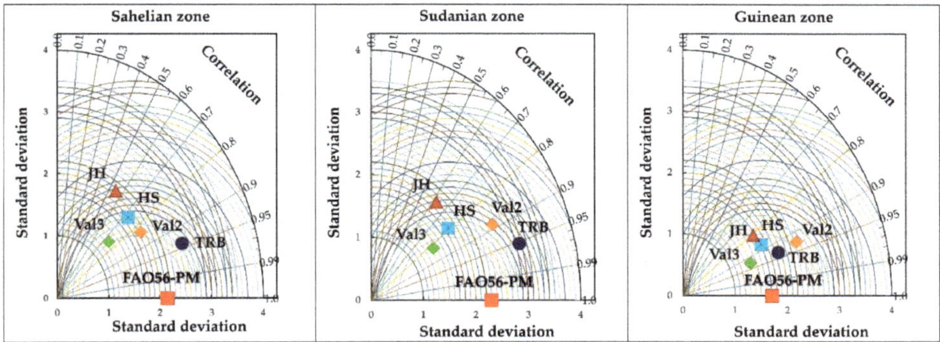

Figure 6. Taylor's diagram of the best methods over the validation period according to climatic zones.

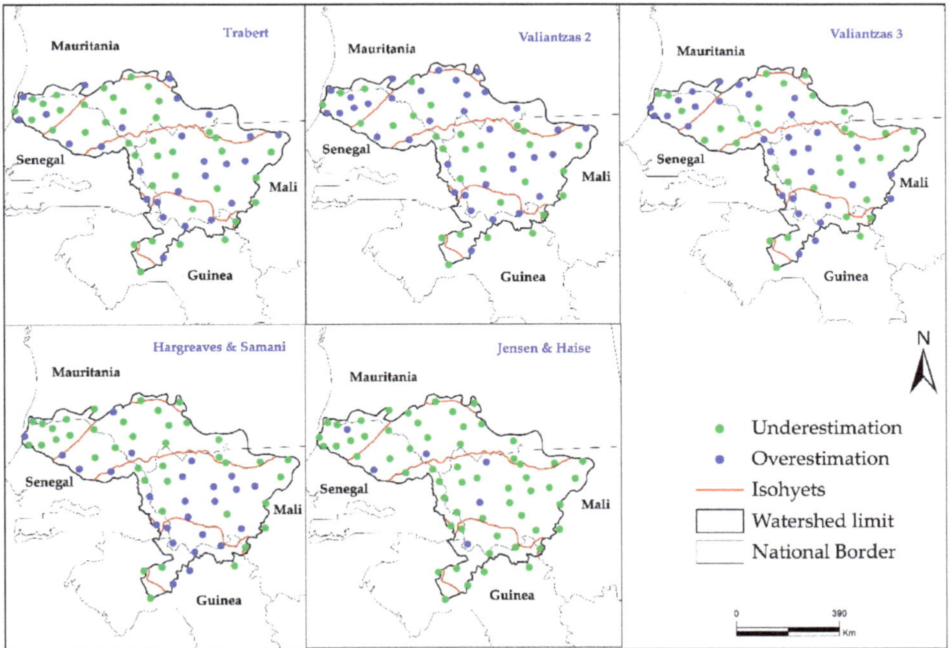

Figure 7. Spatial distribution of percentage of bias (PBIAS) between the best methods and that of FAO56-PM.

4. Conclusions

The objective of this study is to assess the performance of 20 alternative methods for estimating reference evapotranspiration (ET_0) in the Senegal river basin and to calibrate and validate the best methods in order to adapt them to the context of the Senegal river basin. The results show that the Trabert, Valiantzas 2, Valiantzas 3, Hargreaves and Samani, and Jensen and Haise methods are, respectively, the most robust for estimating ET_0 in the Senegal River Basin. Calibration has improved the performance of all these methods except that of Jensen and Haise, whose performance is degraded after calibration. From a spatial point of view, the Trabert method is more efficient in the Sahelian and Sudanian zones. On the other hand, the methods integrating radiation and/or relative humidity

(Valiantzas 2, Valiantzas 3, and Hargreaves and Samani) are more robust in the Guinean area of the basin. This study provides information on the choice of an ET_0 estimation model based on available data and climate zones. When temperature and wind speed data are available, Trabert's method can be used in all climatic zones of the basin for reference evapotranspiration estimation. When relative humidity, radiation, and temperature are available, the Valiantzas 2 method is recommended. The use of the Valiantzas 3 method is only encouraged when radiation and temperature are the only climate variables available. Finally, the HS method can be used when only temperature data are available. These results constitute a source of information on the choice of an adequate model for estimating reference evapotranspiration in the Senegal River Basin. This information can be useful for hydrological modeling, irrigation management, reservoir management, planning, and management of the basin's water resources. However, the types of data (reanalyses) used can cause uncertainties in the results. In addition, the use of a single reanalysis product can also be a source of uncertainty. It would therefore be important to validate these results with in situ data even from a few stations when the availability and accessibility of the information allow.

Author Contributions: P.M.N., A.B., and L.D. designed the study, developed the methodology, and wrote the manuscript. P.M.N. collected the data and conducted the computer analysis with A.B. and L.D., while A.D. (Abdoulaye Deme), A.D. (Alain Dezetter), and K.D. reviewed and edited the paper. All authors have read and agreed to the published version of the manuscript.

Funding: This research has no funding.

Acknowledgments: Data used in this study were obtained from the NASA Langley Research Center (LaRC) POWER Project funded through the NASA Earth Science/Applied Science Program. The authors thank the two reviewers and the editors.

Conflicts of Interest: The authors declare no conflict of interest.

References

1. Xu, C.Y.; Singh, V.P. Evaluation and generalization of temperature-based methods for calculating evaporation. *Hydrol. Process.* **2001**, *14*, 305–319.
2. Oudin, L. Recherche d'un Modèle D'évapotranspiration Potentielle Pertinent Comme Entrée d'un Modèle Pluie-Débit Global (Search for a Relevant Potential Evapotranspiration Model as Input for a Global Rain-Flow Model). Ph.D. Thesis, L'école National de Génie Rural, des Eaux et Des Forêts (ENGREF), Paris, France, 2005; 496p. (In French). Available online: https://pastel.archives-ouvertes.fr/file/index/docid/499816/filename/memoire.pdf (accessed on 15 March 2020).
3. Heydari, M.M.; Aghamajidi, R.; Beygipoor, G.; Heydari, M. Comparison and evaluation of 38 equations for estimating reference evapotranspiration in an arid region. *FEB* **2014**, *23*, 1985–1996.
4. Jian, X.; Scherer, T.; Lin, D.; Zhang, X.; Refali, I. Comparison of reference evapotranspiration calculations for southeastern North Dakota. *Irrig. Drain. Syst. Eng.* **2013**, *2*, 1–9.
5. Birhanu, D.; Kim, H.; Jang, C.; Park, S. Does the complexity of evapotranspiration and hydrological models enhance robustness? *Sustainability* **2018**, *10*, 2837. [CrossRef]
6. Hong, T.; Tran, N.; Honti, M. Application of different evapotranspiration models to calculate total agricultural water demand in a tropical region. *Period. Polytech. Civ. Eng.* **2017**, *61*, 904–910. [CrossRef]
7. Li, Y.; Feng, A.; Liu, W.; Ma, X.; Dong, G. Variation of aridity index and the role of climate variables in the Southwest China. *Water* **2017**, *9*, 743. [CrossRef]
8. Wen, J.; Wang, X.; Guo, M.; Xu, X. Impact of climate change on reference crop evapotranspiration in Chuxiong City, Yunnan Province. *Procedia Earth Planet. Sci.* **2012**, *5*, 113–119.
9. Tao, X.; Hua, C.; Xu, C.; Hou, Y.; Jie, M. Analysis and prediction of reference evapotranspiration with climate change in Xiangjiang River Basin, China. *Water Sci. Eng.* **2015**, *8*, 273–281.
10. Thornthwaite, C.W. An approach toward a rational classification of climate. *Geogr. Rev.* **1948**, *38*, 55–94.

11. Bigeard, G. Estimation Spatialisée de L'évapotranspiration à l'aide de Données Infra-Rouge Thermique Multi-Résolutions (Spatialized Evapotranspiration Estimation Using Multi-Resolution Thermal Infrared Data). Ph.D. Thesis, Université Toulouse III Paul Sabatier (UT3 Paul Sabatier), Toulouse, France, 2014; 259p. (In French). Available online: https://tel.archives-ouvertes.fr/tel-01620222/document (accessed on 15 March 2020).

12. Muhammad, M.K.I.; Nashwan, M.S.; Shahid, S.; Ismail, T.I.; Song, Y.H.; Chung, E. Evaluation of empirical reference evapotranspiration models using compromise programming: A case study of peninsular Malaysia. *Sustainability* **2019**, *11*, 4267. [CrossRef]

13. Jiao, L.; Wang, D. Climate change, the evaporation paradox, and their effects on Streamflow in Lijiang Watershed. *Pol. J. Environ. Stud.* **2018**, *27*, 2585–2591. [CrossRef]

14. Chu, R.; Li, M.; Shen, S.; Islam, A.; Cao, W.; Tao, S.; Gao, P. Changes in reference evapotranspiration and its contributing factors in Jiangsu, a major economic and agricultural province of Eastern China. *Water* **2017**, *9*, 486. [CrossRef]

15. Aubin, A. Estimation de L'évapotranspiration par Télédétection Spatiale en Afrique de L'Ouest: Vers une Meilleure Connaissance de Cette Variable clé Pour la Région (English Title Estimation of Evapotranspiration using Space Remote Sensing in West Africa: Towards a Better Knowledge of this Key Variable for the Region). Ph.D. Thesis, L'Universite Montpellier, Montpellier, France, 2018; 431p. (In French). Available online: https://tel.archives-ouvertes.fr/tel-02045897/document (accessed on 15 March 2020).

16. Cosandey, C.; Robinson, M. *Hydrologie Continentale*; Armand Colin: Paris, France, 2000; 353p.

17. Pereira, L.S.; Allen, R.G.; Smith, M.; Raes, D. Crop evapotranspiration estimation with FAO56: Past and future. *Agric. Water Manag.* **2014**, 1–16. [CrossRef]

18. Allen, R.; Pereira, L.; Raes, D.; Smith, M. *Crop Evapotranspiration. Guideline for Computing Crop Requirements*; FAO-Irrigation and Drainage Paper 56; FAO: Rome, Italy, 1998.

19. Gocic, M.; Trajkovic, S. Analysis of trends in reference evapotranspiration data in a humid climate. *Hydrol. Sci. J.* **2014**, *59*, 165–180.

20. Diop, L.; Bodian, A.; Diallo, D. Use of atmometers to estimate reference evapotranspiration in Arkansas. *Afr. J. Agric. Res.* **2015**, *10*, 4376–4683. [CrossRef]

21. Douar, A. Mesure de L'évapotranspiration par Eddy Covariance. Effet de la Hauteur de Mesure et Variabilité Spatiale (Measuring Evapotranspiration by Eddy Covariance. Effect of Measuring Height and Spatial Variability). Master's Thesis, Université Pierre-Marie-Curie, Paris, France, 2017; 44p. (In French). Available online: http://m2hh.metis.upmc.fr/wp-content/uploads/Douar_Abdelhak_memoireHH1617.pdf (accessed on 11 March 2020).

22. Dalton, J. Experimental essays on the constitution of mixed gases; on the force of steam of vapor from waters and other liquids in different temperatures, both in a Torricellian vacuum and in air on evaporation and on the expansion of gases by heat. *Mem. Manch. Lit. Philos. Soc.* **1802**, *5*, 535–602.

23. Makkink, G.F. Testing the Penman formula by means of lysimeters. *J. Inst. Water Eng.* **1957**, *11*, 277–288.

24. Penman, H.L. *Vegetation and Hydrology*; Technical Communication No. 53; Commonwealth Bureau of Soils: Harpenden, UK, 1963; 125p.

25. Hargreaves, G.H. Moisture availability and crop production. *Trans. ASAE* **1975**, *18*, 980–984.

26. Hargreaves, G.H.; Samani, Z.A. Reference crop evapotranspiration from temperature. *Am. Soc. Agric. Eng.* **1985**, *1*, 96–99.

27. Valiantzas, J. Simple ET0 of Penman's equation without wind/or humidity data. II: Comparisons Reduced Set-FAO and other methodologies. *Am. Soc. Civ. Eng.* **2013**, *139*, 9–19. [CrossRef]

28. Tabari, H. Evaluation of reference crop evapotranspiration equations in various climates. *Water Resour. Manag.* **2010**. [CrossRef]

29. Djaman, K.; Balde, A.B.; Sow, A.; Muller, B.; Irmak, S.; Ndiaye, M.K.; Saito, K. Evaluation of sixteen reference evapotranspiration methods under sahelian conditions in Senegal River Valley. *J. Hydrol. Reg. Stud.* **2015**, *3*, 139–159. [CrossRef]

30. Bodian, A.; Diop, L.; Panthou, G.; Dacosta, H.; Deme, A.; Dezetter, A.; Ndiaye, P.M.; Diouf, I.; Vichel, T. Recent trend in hydroclimatic conditions in the Senegal River Basin. *Water* **2020**, *12*, 436. [CrossRef]

31. Trabert, W. Neue beobachtungen über verdampfungsgeschwindigkeiten (New observations about evaporation rates). *Meteorol. Z.* **1896**, *13*, 261–263. (In German)

32. Penman, H.L. Natural evaporation from open water, bare soil and grass. *Proc. R. Meteorol. Soc.* **1948**, *193*, 120–145.

33. Abtew, W. Evapotranspiration measurement and modeling for three wetland systems in South Florida. *Water Resour. Bull.* **1996**, *32*, 465–473.

34. Rohwer, C. *Evaporation from Free Water Surfaces*; Technical Bulletin 271; US Department of Agriculture: Washington, DC, USA, 1931.

35. Mahringer, W. Verdunstungsstudien am Neusiedler see. *Arch. Meteorol. Geophys. Bioklimatol. Ser. B* **1970**, *18*, 1–20.

36. Trajkovic, S.; Stojvic, V. Effect of wind speed on accuracy of Turc method in humid climate. *Archit. Civ. Eng.* **2007**, *5*, 107–113.

37. Droogers, P.; Allen, R.G. Estimating reference evapotranspiration under inaccurate data conditions. *Irrig. Drain. Syst.* **2002**, *16*, 33–45. [CrossRef]

38. Heydari, M.M.; Heydari, M. Evaluation of pan coefficient equations for estimating reference crop evapotranspiration in the arid region. *Arch. Agron. Soil Sci.* **2014**, *60*, 715–731. [CrossRef]

39. Jensen, M.E.; Haise, H.R. Estimating evapotranspiration from solar radiation. *J. Irrig. Drain. Div.* **1963**, *89*, 15–41.

40. Priestley, C.H.B.; Taylor, R.J. On the assessment of surface heat flux and evaporation using large scale parameters. *Mon. Weath. Rev.* **1972**, *100*, 81–92.

41. Doorenbos, J.; Pruitt, W.O. *Guidelines for Predicting Crop Water Requirements*; FAO Irrigation and Drainag, Paper No. 24; FAO: Rome, Italy, 1977.

42. Xu, C.Y.; Singh, V.P. Evaluation and generalization of radiation-based methods for calculating evaporation. *Hydrol. Process.* **2000**, *14*, 339–349.

43. Ahooghalandari, M.; Khiadani, M.; Jahromi, M.E. Calibration of Valiantzas' reference evapotranspiration equations for the Pilbara region, Western Australia. *Appl. Clim.* **2016**. [CrossRef]

44. Čadro, S.; Uzunovi, M.; Žurovec, J.; Žurovec, O. Validation and calibration of various reference evapotranspiration alternative methods under the climate conditions of Bosnia and Herzegovina. *Int. Soil Water Conserv. Res.* **2017**, *5*, 309–324. [CrossRef]

45. Djaman, K.; Tabari, H.; Balde, A.B.; Diop, L.; Futakuchi, K.; et Irmak, S. Analyses, calibration and validation of evapotranspiration models to predict grass-reference evapotranspiration in the Senegal river delta. *J. Hydrol. Reg. Stud.* **2016**, *8*, 82–94. [CrossRef]

46. Bodian, A. Approche par Modélisation Pluie-Débit de la Connaissance Régionale de la Ressource en Eau: Application dans le Haut Bassin du Fleuve Sénégal (Rain-Flow Modeling Approach to Regional Knowledge of Water Resources: Application in the Upper Basin of the Senegal River). Ph.D. Thesis, Université Cheikh Anta Diop de Dakar, Dakar, Senegal, 2011; 211p. (In French). Available online: http://hydrologie.org/THE/BODIAN.pdf (accessed on 10 March 2020).

47. Dione, O. Evolution Climatique Récente et Dynamique Fluviale dans les Hauts Bassins des Fleuves Sénégal et Gambie (Recent Climate Evolution and Fluvial Dynamics in the High Basins of the Senegal and Gambia Rivers). Ph.D. Thesis, Université de Lyon 3 Jean Moulin, ORSTOM, Paris, France, 1996; 438p. (In French). Available online: http://horizon.documentation.ird.fr/exl-doc/pleins_textes/pleins_textes_7/TDM_7/010012551.pdf (accessed on 10 March 2020).

48. SDAGE-OMVS. *Etat des Lieux et Diagnostique*; Rapport Provisoire 2009, Rapport de Phase III; SDAGE-OMVS: Dakar, Senegal, 2011.

49. Bodian, A.; Dezetter, A.; Deme, A.; Diop, L. Hydrological evaluation of TRMM rainfall over the upper Senegal River Basin. *Hydrology* **2016**, *3*, 15. [CrossRef]

50. Srivastava, P.; Han, D.; Ramirez, M.A.; Islam, T. Comparative assessment of evapotranspiration derived from NCEP and ECMWF global datasets through Weather Research and Forecasting model. *Atmos. Sci. Lett.* **2013**, *14*, 118–125. [CrossRef]

51. Poccard-Leclercq, I. Etude Diagnostique de Nouvelles Données Climatiques: Les Réanalyses. Exemples D'application aux Précipitations en Afrique Tropicale (Diagnostic Study of New Climate Data: Reanalyses. Application to Precipitation in Tropical Africa). PhD. Thesis, Géographie. Université de Bourgogne, Dijon, France, 2000; 255p. (In French)

52. Ruane, A.C.; Goldberg, R.; Chryssanthacopoulos, J. Climate forcing datasets for agricultural modeling: Merged products for gap-filling and historical climate series estimation. *Agric. For. Meteorol.* **2015**, *200*, 233–248.

53. Martins, D.S.; Paredes, P.; Razia, T.; Pires, C.; Cadima, J.; Pereira, L. Assessing reference evapotranspiration from reanalysis weather products. An application to the Iberian Peninsula. *Int. J. Climatol.* **2016**, *37*, 1–20. [CrossRef]

54. Stackhouse, P.W.; Westberg, D., Jr.; Chandler, W.S.; Zhang, T.; Hoell, J.M. Prediction of Worldwide Energy Resource (POWER): Agroclimatology Methodology. 2018; 52p. Available online: https://power.larc.nasa.gov/documents/POWER_Data_v9_methodology.pdf (accessed on 20 December 2018).

55. Purnadurga, G.T.V.; Kumar, L.; Rao, K.K.; Barbosa, H.; Mall, R.K. Evaluation of evapotranspiration estimates from observed and reanalysis data sets over Indian region. *Int. J. Climatol.* **2019**, *39*, 5791–5800. [CrossRef]

56. Gupta, H.V.; Kling, H.; Yilmaz, K.K.; Martinez, G.F. Decomposition of the mean squared error and NSE performance criteria: Implications for improving hydrological modeling. *J. Hydrol.* **2009**, *377*, 80–91.

57. Valipour, M. Calibration of mass transfer-based methods to predict reference crop evapotranspiration. *Appl. Water Sci.* **2015**, 1–11. [CrossRef]

58. Bogawski, P.; Bednorz, E. Comparison and validation of selected evapotranspiration models for conditions in Poland (Central Europe). *Water Resour. Manag.* **2014**, *28*, 5021–5038. [CrossRef]

59. Taylor, K.E. Summarizing multiple aspects of model performance in a single diagram. *J. Geophys. Res. Atmos.* **2011**, *106*, 7183–7192. [CrossRef]

60. Singh, V.P.; Xu, C.Y. Dependence of evaporation on meteorological variables at different time scales and intercomparison of estimation methods. *Hydrol. Process.* **1997**, *12*, 429–442.

61. Djaman, K.; Koudahe, K.; Sall, M.; Kabenge, I.; Rudnick, D.; Irmak, S. Performance of twelve mass transfer based reference evapotranspiration models under humid climate. *J. Water Resour. Prot.* **2017**, *9*, 1347–1363. [CrossRef]

62. Ndiaye, P.M.; Bodian, A.; Diop, L.; Djaman, K. Evaluation de vingt méthodes d'estimation de l'évapotranspiration journalière de référence au Burkina Faso. *Physio-Géo* **2017**, *11*, 129–146.

63. Tabari, H.; Talaee, P.H. Sensitivity of evapotranspiration to climatic change in different climates. *Glob. Planet. Chang.* **2014**, *115*, 16–23.

64. Irmak, S.; Allen, R.G.; Whitty, E.B. Daily grass and alfalfa-reference evapotranspiration estimates and alfalfa-to-grass evapotranspiration ratios in Florida. *J. Irrig. Drain. Eng.* **2003**, *129*, 360–370. [CrossRef]

65. Ambas, V.T.; Baltas, E. Sensitivity analysis of different evapotranspiration methods using a new sensitivity coefficient. *Glob. Nest J.* **2012**, *14*, 335–343.

MDPI

St. Alban-Anlage 66

4052 Basel

Switzerland

Tel. +41 61 683 77 34

Fax +41 61 302 89 18

www.mdpi.com

Hydrology Editorial Office

E-mail: hydrology@mdpi.com

www.mdpi.com/journal/hydrology